1st Tensor Polarized Solid Target Workshop 2014

AA008920

Journal of Physics: Conference Series Volume 543

Newport News, Virginia, USA
10-12 March 2014

ISBN: 978-1-63439-594-6
ISSN: 1742-6588

Printed from e-media with permission by:

Curran Associates, Inc.
57 Morehouse Lane
Red Hook, NY 12571

proceedings.com

Some format issues inherent in the e-media version may also appear in this print version.

Printed by Curran Associates, Inc. (2015)

For permission requests, please contact the Institute of Physics
at the address below.

Institute of Physics
Dirac House, Temple Back
Bristol BS1 6BE UK

Phone: 44 1 17 929 7481
Fax: 44 1 17 920 0979

techtracking@iop.org

Additional copies of this publication are available from:

Curran Associates, Inc.
57 Morehouse Lane
Red Hook, NY 12571 USA
Phone: 845-758-0400
Fax: 845-758-2634
Email: curran@proceedings.com
Web: www.proceedings.com

TABLE OF CONTENTS

THEORY AND EXPERIMENT

IDEAS FOR NEW TENSOR – POLARIZED EXPERIMENTS

SOLID TENSOR – POLARIZED TARGETS

Journal of Physics: Conference Series **543** (2014) 011001 doi:10.1088/1742-6596/543/1/011001

Preface

These are the proceedings of the first Tensor Spin Observables Workshop that was held in March 2014 at the Thomas Jefferson National Accelerator Facility in Newport News, Virginia. The conference was convened to study the physics that can be done with the recently approved E12-13-011 polarized target. A tensor polarized target holds the potential of initiating a new generation of tensor spin physics at Jefferson Lab. Experiments which utilize tensor polarized targets can help clarify how nuclear properties arise from partonic degrees of freedom, provide unique insight into short-range correlations and quark angular momentum, and also help pin down the polarization of the quark sea with a future Electron Ion Collider.

This three day workshop was focused on tensor spin observables and the associated tensor target development. The workshop goals were to stimulate progress in the theoretical treatment of polarized spin-1 systems, foster the development of new proposals, and to reach a consensus on the optimal polarized target configuration for the tensor spin program. The workshop was sponsored by the University of New Hampshire, the Jefferson Science Associates, Florida International University, and Jefferson Lab. It was organized by Karl Slifer (chair), Patricia Solvignon, and Elena Long of the University of New Hampshire, Douglas Higinbotham and Christopher Keith of Jefferson Lab, and Misak Sargsian of the Florida International University. These proceedings represent the effort put forth by the community to begin exploring the possibilities that a high-luminosity, high-tensor polarized solid target can offer.

Elena Long (University of New Hampshire)
Douglas Higinbotham (Jefferson Lab)
Patricia Solvignon (University of New Hampshire)

Journal of Physics: Conference Series **543** (2014) 011001 doi:10.1088/1742-6596/543/1/011001

Peer review statement

All papers published in this volume of *Journal of Physics: Conference Series* have been peer reviewed through processes administered by the proceedings Editors. Reviews were conducted by expert referees to the professional and scientific standards expected of a proceedings journal published by IOP Publishing.

Tensor-polarized structure functions: Tensor structure of deuteron in 2020's

S Kumano

KEK Theory Center, Institute of Particle and Nuclear Studies, KEK
1-1, Ooho, Tsukuba, Ibaraki, 305-0801, Japan
J-PARC Branch, KEK Theory Center, Institute of Particle and Nuclear Studies, KEK and Theory Group, Particle and Nuclear Physics Division, J-PARC Center
203-1, Shirakata, Tokai, Ibaraki, 319-1106, Japan

E-mail: shunzo.kumano@kek.jp

Abstract. We explain spin structure for a spin-one hadron, in which there are new structure functions, in addition to the ones (F_1, F_2, g_1, g_2) which exist for the spin-1/2 nucleon, associated with its tensor structure. The new structure functions are b_1, b_2, b_3, and b_4 in deep inelastic scattering of a charged-lepton from a spin-one hadron such as the deuteron. Among them, twist-two functions are related by the Callan-Gross type relation $b_2 = 2xb_1$ in the Bjorken scaling limit. First, these new structure functions are introduced, and useful formulae are derived for projection operators of b_{1-4} from a hadron tensor $W_{\mu\nu}$. Second, a sum rule is explained for b_1, and possible tensor-polarized distributions are discussed by using HERMES data in order to propose future experimental measurements and to compare them with theoretical models. A proposal was approved to measure b_1 at the Thomas Jefferson National Accelerator Facility (JLab), so that much progress is expected for b_1 in the near future. Third, formalisms of polarized proton-deuteron Drell-Yan processes are explained for probing especially tensor-polarized antiquark distributions, which were suggested by the HERMES data. The studies of the tensor-polarized structure functions will open a new era in 2020's for tensor-structure studies in terms of quark and gluon degrees of freedom, which are very different from ordinary descriptions in terms of nucleons and mesons.

1. Introduction

Spin structure of the nucleon has been extensively investigated for finding its origin. In a naive quark model, the spin-1/2 nucleon should consist of two quarks with parallel spins to the nucleon spin and a quark with antiparallel spin. In other words, quarks carry 100% of the nucleon spin. However, polarized lepton-nucleon scattering measurements clarified that this simple description does not work. The quarks carry only a small fraction of the nucleon spin. The remaining spin may be possibly carried by antiquark and gluon spins. However, antiquark spin contributions are small and gluon spin contribution seems to be also small according to recent measurements and their global analyses. Therefore, only remaining possibilities are orbital angular momentum effects of quarks and gluons. They are currently investigated by generalized parton distributions (GPDs) and transverse-momentum-dependent parton distributions (TMDs). Therefore, understanding of orbital motion is crucial in hadron spin physics.

Journal of Physics: Conference Series **543** (2014) 012001 doi:10.1088/1742-6596/543/1/012001

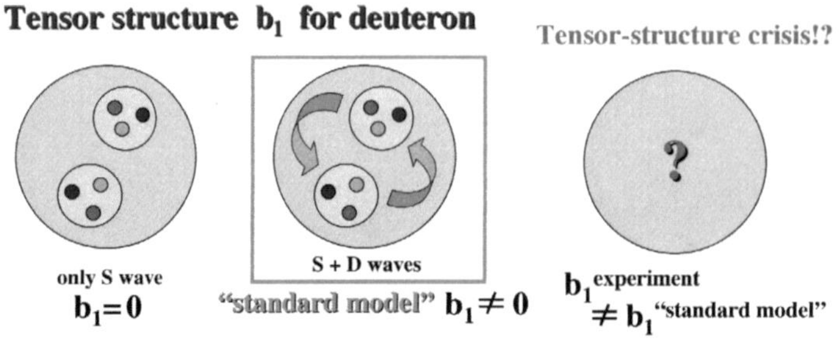

Figure 1. Tensor structure of the deuteron

For a spin-one hadron such as the deuteron, there exist new polarized structure functions, b_1, b_2, b_3, and b_4, which could probe dynamical aspects including orbital motion within the hadron. The new structure is associated with tensor structure, so that the tensor-polarized structure functions vanish if constituents are in the S state. It is also important to investigate completely different spin quantities from the nucleon for establishing high-energy spin physics.

The tensor structure of deuteron and other nuclei has been investigated for a long time by nucleon and meson degrees of freedom, as illustrated in Fig. 1. However, the tensor structure has not been explored yet in terms of quark and gluon degrees of freedom. There are some studies. First, the new structure functions b_{1-4} are introduced for a spin-one hadron [1, 2]. For the leading-twist structure function b_1, a useful sum rule was proposed by using a parton model [3]. Leptoproduction of spin-one hadrons is investigated in Ref. [4]. There are theoretical studies on the x distribution of b_1. If the constituents are in the S wave, it vanishes ($b_1 = 0$) [2]. A finite distribution with a node is expected in a convolution model with D-state admixture [2, 5], effects of pions and six-quark configuration [6], and shadowing effects in a nucleus [7, 8]. If we find a significant difference from these theoretical expectations, the studies could create a new field of high-energy spin physics. There are related theoretical studies such as new fragmentation functions [9], generalized parton distributions [10], target mass corrections [11], positivity constraints [12], lattice QCD estimate [13], projection operators of b_{1-4} [14] for spin-one hadrons, and angular momenta for spin-1 hadron [15]. An alternative reaction for investigating the tensor structure is to use Drell-Yan processes. A theoretical formalism was developed in Ref. [16] to study the tensor-polarized distributions at hadron facilities by Drell-Yan processes with polarized deuteron.

The HERMES collaboration reported the measurement of the structure function b_1 in 2005 [17]. The data indicated a finite distribution at $x < 0.1$, which roughly agrees with a double scattering contribution estimated in Ref. [8]. The data are also consistent with the quark-parton model sum rule for b_1 [3] although experimental errors are still large. As a future experiment, a proposal was approved for measuring b_1 at Thomas Jefferson National Accelerator Facility (JLab) [18]. It could be also investigated at the Electron-Ion Collider (EIC) [19]. Tensor polarization A_{zz} at large-x ($x > 1$) can be also investigated [20]. In addition, the tensor-polarized quark and antiquark distributions could be studied by Drell-Yan processes with tensor-polarized deuteron at Fermilab [21], J-PARC (Japan Proton Accelerator Research Complex) [22], and GSI-FAIR (Gesellschaft für Schwerionenforschung -Facility for Antiproton and Ion Research) [23].

In this article, we explain the definition of tensor-polarized structure functions in Sec. 2. Then, operators to project out structure functions of a spin-1 hadron are shown in Sec. 3. A sum rule of $b_1(x)$ is explained in Sec. 4. A useful parametrization is proposed for tensor-polarized distributions in Sec. 5 for explaining the HERMES data. As shown in Sec. 6, the tensor-polarized distributions could be also investigated by Drell-Yan processes with tensor-polarized deuteron [16]. The results are summarized in Sec. 7.

Journal of Physics: Conference Series **543** (2014) 012001 doi:10.1088/1742-6596/543/1/012001

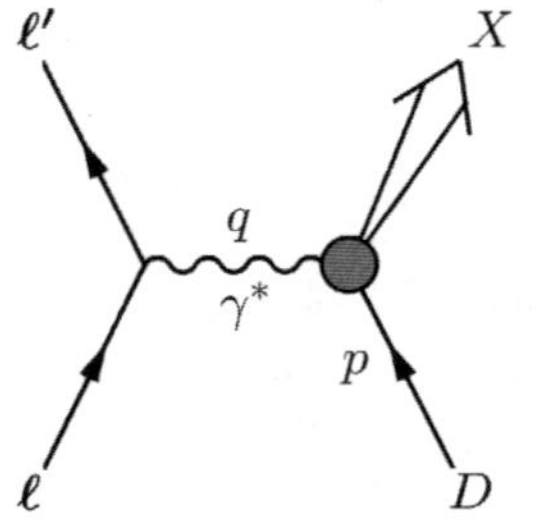

Figure 2. Deep inelastic scattering of charged lepton from a hadron D.

2. Tensor polarized structure functions

We consider charged-lepton deep inelastic scattering (DIS) from a hadron in Fig. 2. Its cross section is described by a hadron tensor $W_{\mu\nu}$ multiplied by a lepton tensor $L^{\mu\nu}$. The hadron tensor for the spin-1/2 nucleon is described by four structure functions F_1, F_2, g_1, and g_2:

$$W_{\mu\nu}^{\lambda_f \lambda_i} = -F_1 \hat{g}_{\mu\nu} + \frac{F_2}{M\nu}\hat{p}_\mu \hat{p}_\nu + \frac{ig_1}{\nu}\epsilon_{\mu\nu\lambda\sigma}q^\lambda s^\sigma + \frac{ig_2}{M\nu^2}\epsilon_{\mu\nu\lambda\sigma}q^\lambda(p\cdot q s^\sigma - s\cdot q p^\sigma), \tag{1}$$

$$\hat{g}_{\mu\nu} \equiv g_{\mu\nu} - \frac{q_\mu q_\nu}{q^2}, \quad \hat{a}_\mu \equiv a_\mu - \frac{a\cdot q}{q^2}q_\mu, \tag{2}$$

where $\epsilon_{\mu\nu\lambda\sigma}$ is an antisymmetric tensor with the convention $\epsilon_{0123} = +1$, ν is defined by $\nu = p\cdot q/M$ with the hadron mass M, hadron momentum p, and momentum transfer q, Q^2 is given by $Q^2 = -q^2 > 0$, and s^μ is the spin vector with the constraint $s\cdot p = 0$. The notations $\hat{g}_{\mu\nu}$ and $\hat{p}_\mu$ are introduced so as to ensure the current conservation $q^\mu W_{\mu\nu} = q^\nu W_{\mu\nu} = 0$. The initial and final spin states are denoted by λ_i and λ_f, respectively. In general, off-diagonal terms with $\lambda_f \neq \lambda_i$ are needed to consider higher-twist contributions [2].

In a spin-one hadron, there are four additional structure functions b_1, b_2, b_3 and b_4 in the hadron tensor [2, 14]:

$$\begin{aligned} W_{\mu\nu}^{\lambda_f \lambda_i} =& - F_1 \hat{g}_{\mu\nu} + \frac{F_2}{M\nu}\hat{p}_\mu \hat{p}_\nu + \frac{ig_1}{\nu}\epsilon_{\mu\nu\lambda\sigma}q^\lambda s^\sigma + \frac{ig_2}{M\nu^2}\epsilon_{\mu\nu\lambda\sigma}q^\lambda(p\cdot q s^\sigma - s\cdot q p^\sigma) \\ & - b_1 r_{\mu\nu} + \frac{1}{6}b_2(s_{\mu\nu} + t_{\mu\nu} + u_{\mu\nu}) + \frac{1}{2}b_3(s_{\mu\nu} - u_{\mu\nu}) + \frac{1}{2}b_4(s_{\mu\nu} - t_{\mu\nu}), \end{aligned} \tag{3}$$

where $r_{\mu\nu}$, $s_{\mu\nu}$, $t_{\mu\nu}$, and $u_{\mu\nu}$ are defined by

$$\begin{aligned} r_{\mu\nu} =& \frac{1}{\nu^2}\left[q\cdot E^*(\lambda_f) q\cdot E(\lambda_i) - \frac{1}{3}\nu^2\kappa\right]\hat{g}_{\mu\nu}, \qquad s_{\mu\nu} = \frac{2}{\nu^2}\left[q\cdot E^*(\lambda_f) q\cdot E(\lambda_i) - \frac{1}{3}\nu^2\kappa\right]\frac{\hat{p}_\mu\hat{p}_\nu}{M\nu}, \\ t_{\mu\nu} =& \frac{1}{2\nu^2}\left[q\cdot E^*(\lambda_f)\left\{\hat{p}_\mu \hat{E}_\nu(\lambda_i) + \hat{p}_\nu \hat{E}_\mu(\lambda_i)\right\} + \left\{\hat{p}_\mu \hat{E}^*_\nu(\lambda_f) + \hat{p}_\nu \hat{E}^*_\mu(\lambda_f)\right\} q\cdot E(\lambda_i) - \frac{4\nu}{3M}\hat{p}_\mu\hat{p}_\nu\right], \\ u_{\mu\nu} =& \frac{M}{\nu}\left[\hat{E}^*_\mu(\lambda_f)\hat{E}_\nu(\lambda_i) + \hat{E}^*_\nu(\lambda_f)\hat{E}_\mu(\lambda_i) + \frac{2}{3}\hat{g}_{\mu\nu} - \frac{2}{3M^2}\hat{p}_\mu\hat{p}_\nu\right]. \end{aligned} \tag{4}$$

Here, κ is given by $\kappa = 1 + Q^2/\nu^2$, and s^μ is the spin vector of the spin-one hadron. The E^μ is the polarization vector of the spin-one hadron and it satisfies the conditions, $p\cdot E = 0$ and $E^*\cdot E = -1$. It is taken as the spherical unit vectors:

$$E^\mu(\lambda = \pm 1) = \frac{1}{\sqrt{2}}(0, \mp 1, -i, 0), \quad E^\mu(\lambda = 0) = (0,0,0,1). \tag{5}$$

Then, the spin vector s is given by the polarization vector as

$$(s_{\lambda_f \lambda_i})^\mu = -\frac{i}{M}\epsilon^{\mu\nu\alpha\beta} E_\nu^*(\lambda_f) E_\alpha(\lambda_i) p_\beta, \tag{6}$$

where M is the mass of the spin-1 hadron. The initial and final polarization vectors are denoted by $E^\mu(\lambda_i)$ and $E^\mu(\lambda_f)$, respectively, with the spin states λ_i and λ_f. Explicit expressions for the spin vector s^μ are given for s^μ_{11}, s^μ_{10}, and s^μ_{01} as

$$s^\mu_{11} = (0,0,0,1), \quad s^\mu_{10} = \frac{1}{\sqrt{2}}(0,1,-i,0), \quad s^\mu_{01} = (s^\mu_{10})^*. \tag{7}$$

In Eq. (3), the coefficients of b_1, b_2, b_3 and b_4 are symmetric under $\mu \leftrightarrow \nu$, and they vanish under the spin average. The coefficients are defined so that b_1 and b_2 are twist-two functions and they satisfy the Callan-Gross type relation. In addition to the functions F_1, F_2, g_1, and g_2, which exist in a spin-1/2 hadron, there are four new structure functions, b_1, b_2, b_3 and b_4. They are related to the tensor structure of the spin-1/2 hadron.

3. Projection operators for tensor-polarized structure functions

In a convolution description of nuclear structure functions, the hadron tensor $W_{\mu\nu}$ is calculated by the convolution integral of nucleonic $W_{\mu\nu}$ with a lightcone momentum distribution of a nucleon in a nucleus. Then, the nuclear structure function F_2^A is, for example, extracted from the nuclear tensor $W^A_{\mu\nu}$ by applying a projection operator to $W^A_{\mu\nu}$. Because there are eight structure functions for a spin-one hadron, eight independent combinations are needed for tensors with indices μ and ν. Only $\lambda = 1$ and $\lambda = 0$ terms are used because $\lambda = -1$ terms make the same contributions as $\lambda = 1$ ones. We choose the following terms

$$g^{\mu\nu}\delta_{\lambda_f 1}\delta_{\lambda_i 1}, \quad g^{\mu\nu}\delta_{\lambda_f 0}\delta_{\lambda_i 0}, \quad g^{\mu\nu}\delta_{\lambda_f 1}\delta_{\lambda_i 0}, \quad \frac{p^\mu p^\nu}{M^2}\delta_{\lambda_f 1}\delta_{\lambda_i 1}, \quad \frac{p^\mu p^\nu}{M^2}\delta_{\lambda_f 0}\delta_{\lambda_i 0},$$
$$\frac{1}{M}[p^\mu E^\nu(\lambda=1) + p^\nu E^\mu(\lambda=1)]\,\delta_{\lambda_f 1}\delta_{\lambda_i 0}, \quad \frac{i}{M}\epsilon^{\mu\nu\alpha\beta} q_\alpha s^{11}_\beta \delta_{\lambda_f 1}\delta_{\lambda_i 1}, \quad \frac{i}{M}\epsilon^{\mu\nu\alpha\beta} q_\alpha s^{10}_\beta \delta_{\lambda_f 0}\delta_{\lambda_i 1}. \tag{8}$$

Then, we obtain the projection operators for the structure functions of a spin-one hadron as [14]

$$
\begin{aligned}
F_1 &= -\frac{1}{2}\left(g^{\mu\nu} - \frac{\kappa-1}{\kappa}\frac{p^\mu p^\nu}{M^2}\right)\frac{1}{3}\delta_{\lambda_f\lambda_i}W_{\mu\nu}^{\lambda_f\lambda_i}, \\
F_2 &= -\frac{x}{\kappa}\left(g^{\mu\nu} - \frac{\kappa-1}{\kappa}\frac{3p^\mu p^\nu}{M^2}\right)\frac{1}{3}\delta_{\lambda_f\lambda_i}W_{\mu\nu}^{\lambda_f\lambda_i}, \\
g_1 &= -\frac{i}{2\kappa\nu}\epsilon^{\mu\nu\alpha\beta}q_\alpha\left(s_\beta^{11}\delta_{\lambda_f 1}\delta_{\lambda_i 1} - s_\beta^{10}\delta_{\lambda_f 0}\delta_{\lambda_i 1}\right)W_{\mu\nu}^{\lambda_f\lambda_i}, \\
g_2 &= \frac{i}{2\kappa\nu}\epsilon^{\mu\nu\alpha\beta}q_\alpha\left(s_\beta^{11}\delta_{\lambda_f 1}\delta_{\lambda_i 1} + \frac{s_\beta^{10}}{\kappa-1}\delta_{\lambda_f 0}\delta_{\lambda_i 1}\right)W_{\mu\nu}^{\lambda_f\lambda_i}, \\
b_1 &= \left[-\frac{1}{2\kappa}g^{\mu\nu}(\delta_{\lambda_f 0}\delta_{\lambda_i 0} - \delta_{\lambda_f 1}\delta_{\lambda_i 0}) + \frac{\kappa-1}{2\kappa^2}\frac{p^\mu p^\nu}{M^2}(\delta_{\lambda_f 0}\delta_{\lambda_i 0} - \delta_{\lambda_f 1}\delta_{\lambda_i 1})\right]W_{\mu\nu}^{\lambda_f\lambda_i}, \\
b_2 &= \frac{x}{\kappa^2}\left[g^{\mu\nu}\left\{-\delta_{\lambda_f 0}\delta_{\lambda_i 0} - 2(\kappa-1)\delta_{\lambda_f 1}\delta_{\lambda_i 1} + (2\kappa-1)\delta_{\lambda_f 1}\delta_{\lambda_i 0}\right\}\right. \\
&\left. + \frac{3(\kappa-1)}{\kappa}\frac{p^\mu p^\nu}{M^2}(\delta_{\lambda_f 0}\delta_{\lambda_i 0} - \delta_{\lambda_f 1}\delta_{\lambda_i 1}) - \frac{4(\kappa-1)}{\sqrt{\kappa}M}\{p^\mu E^\nu(\lambda=1) + p^\nu E^\mu(\lambda=1)\}\delta_{\lambda_f 1}\delta_{\lambda_i 0}\right]W_{\mu\nu}^{\lambda_f\lambda_i}, \\
b_3 &= \frac{x}{3\kappa^2}\left[g^{\mu\nu}\left\{-\delta_{\lambda_f 0}\delta_{\lambda_i 0} + \frac{2(2\kappa^2+2\kappa-1)}{\kappa-1}\delta_{\lambda_f 1}\delta_{\lambda_i 1} - \frac{4\kappa^2+3\kappa-1}{\kappa-1}\delta_{\lambda_f 1}\delta_{\lambda_i 0}\right\}\right. \\
&\left. + \frac{3(\kappa-1)}{\kappa}\frac{p^\mu p^\nu}{M^2}(\delta_{\lambda_f 0}\delta_{\lambda_i 0} - \delta_{\lambda_f 1}\delta_{\lambda_i 1}) - \frac{4(\kappa-1)}{\sqrt{\kappa}M}\{p^\mu E^\nu(\lambda=1) + p^\nu E^\mu(\lambda=1)\}\delta_{\lambda_f 1}\delta_{\lambda_i 0}\right]W_{\mu\nu}^{\lambda_f\lambda_i}, \\
b_4 &= \frac{x}{3\kappa^2}\left[g^{\mu\nu}\left\{-\delta_{\lambda_f 0}\delta_{\lambda_i 0} - \frac{2(\kappa^2+4\kappa+1)}{\kappa-1}\delta_{\lambda_f 1}\delta_{\lambda_i 1} + \frac{2\kappa^2+9\kappa+1}{\kappa-1}\delta_{\lambda_f 1}\delta_{\lambda_i 0}\right\}\right. \\
&\left. + \frac{3(\kappa-1)}{\kappa}\frac{p^\mu p^\nu}{M^2}(\delta_{\lambda_f 0}\delta_{\lambda_i 0} - \delta_{\lambda_f 1}\delta_{\lambda_i 1}) + \frac{4(2\kappa+1)}{\sqrt{\kappa}M}\{p^\mu E^\nu(\lambda=1) + p^\nu E^\mu(\lambda=1)\}\delta_{\lambda_f 1}\delta_{\lambda_i 0}\right]W_{\mu\nu}^{\lambda_f\lambda_i},
\end{aligned}
\tag{9}
$$

where summations are taken over λ_i and λ_f.

Because leading-twist structure functions b_1 and b_2 were experimentally measured first, it is useful to consider the Bjorken scaling limit, ν, $Q^2 \to \infty$ with finite $x = Q^2/(2p\cdot q)$. In this limit, we have the relations

$$
\begin{aligned}
&\lim_{\rm Bj} g^{\mu\nu}W_{\mu\nu}^{10} = \lim_{\rm Bj} g^{\mu\nu}W_{\mu\nu}^{11}, \quad \lim_{\rm Bj}(\kappa-1)\frac{p^\mu p^\nu}{M^2}W_{\mu\nu}^{\lambda\lambda} = 0, \\
&\lim_{\rm Bj}\frac{\kappa-1}{M}[p^\mu E^\nu(\lambda=1) + p^\nu E^\mu(\lambda=1)]W_{\mu\nu}^{10} = 0,
\end{aligned}
\tag{10}
$$

by noting $\kappa \to 1$, $2xF_1 \to F_2$, and $2xb_1 \to b_2$. Then we obtain the expressions from the projection operator in the Bjorken scaling limit as

$$
\begin{aligned}
F_1 &= \frac{1}{2x}F_2 = -\frac{1}{2}g^{\mu\nu}\frac{1}{3}\delta_{\lambda_f\lambda_i}W_{\mu\nu}^{\lambda_f\lambda_i}, \quad g_1 = -\frac{i}{2\nu}\epsilon^{\mu\nu\alpha\beta}q_\alpha s_\beta^{11}\delta_{\lambda_f 1}\delta_{\lambda_i 1}W_{\mu\nu}^{\lambda_f\lambda_i}, \\
b_1 &= \frac{1}{2x}b_2 = \frac{1}{2}g^{\mu\nu}(\delta_{\lambda_f 1}\delta_{\lambda_i 1} - \delta_{\lambda_f 0}\delta_{\lambda_i 0})W_{\mu\nu}^{\lambda_f\lambda_i}.
\end{aligned}
\tag{11}
$$

The exact relations of Eq. (9) or the ones of Eq. (11) in the scaling limit are useful equations for extracting the new structure functions b_{1-4} in addition to the usual ones, $F_{1,2}$ and $g_{1,2}$, for estimating them in theoretical descriptions, especially in the convolution model.

Journal of Physics: Conference Series **543** (2014) 012001 doi:10.1088/1742-6596/543/1/012001

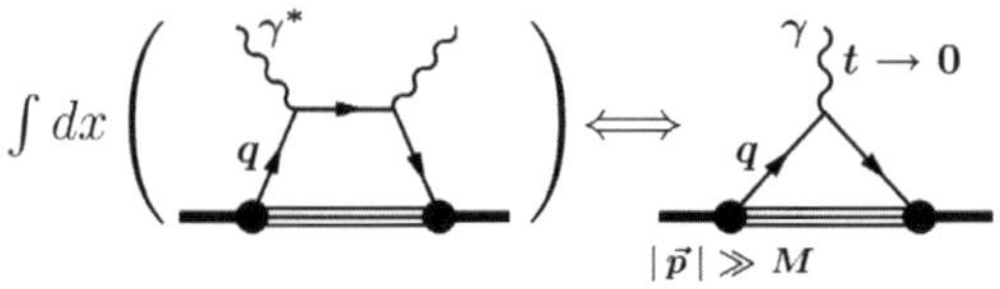

Figure 3. Structure function and elastic form factor by the parton model in the frame with $|\vec{p}| \gg M$.

4. Sum rule for b_1

We discuss a useful sum rule for the structure function b_1. Sum rules of structure functions could be derived in a parton model by considering the relation between the structure function integrated over x and elastic form factor in the infinite momentum frame [24]. In the parton model, the structure function b_1 is expressed by the tensor-polarized distributions $\delta_T q(x)$ as [2, 3][1]

$$b_1(x, Q^2) = \frac{1}{2} \sum_i e_i^2 \left[\delta_T q_i(x, Q^2) + \delta_T \bar{q}_i(x, Q^2) \right], \quad \delta_T q_i \equiv q_i^0 - \frac{q_i^{+1} + q_i^{-1}}{2}, \tag{12}$$

where i indicates the flavor of a quark and e_i is its charge, and q_i^λ indicates an unpolarized-quark distribution in the hadron spin state λ. Instead of δ_T, δ could be used for the tensor-polarized distributions. Because δ and Δ_T are often used for transversity distributions of spin-1/2 nucleon, the notation δ_T is used throughout this article. The function $\delta_T q$ indicates an unpolarized-quark distribution in a tensor-polarized spin-one hadron. The idea how to derive a sum rule for b_1 is illustrated in Fig. 3. The hadron tensor is equal to the imaginary part of the forward scattering amplitude for virtual photon scattering from the hadron by the optical theorem. An integral of a DIS structure function is related to the form factor in the infinite momentum frame by expressing the form factor in terms of parton momentum distributions as discussed in Ref. [24], where the Bjorken sum rule is shown by this method.

Before stepping into the details, we may estimate the sum rule of $b_1(x)$ intuitively by using the dimensional counting. The b_1 is defined in Eq. (3) as a dimensionless quantity. If it is integrated over x, it should become electromagnetic quantities of the deuteron probed by the charged lepton. Due to the parity conservation, the new quantity, which does not exist in the spin-1/2 nucleon, should be the electric quadrupole moment. The electric quadrupole moment has the dimension of length^2=1/mass^2 except for the charge factor. Therefore, the b_1 sum is expected to be $\int dx b_1(x) \sim (\text{mass dim.})^2 \cdot Q_h$, where Q_h is the electric quadrupole moment of the hadron h. The quantity of the mass dimension may be the deuteron mass M, but it is not obvious, so we proceed to a more detailed calculation.

For the deuteron, we define the structure function b_1 and parton distribution functions (PDFs) by the ones per nucleon ($b_1/2 \to b_1$, $q(x)/2 \to q(x)$) Taking the integral of Eq. (12) over x and then using the relations, $(\delta_T u_v)_D \equiv (\delta_T u - \delta_T \bar{u})_D = (\delta_T u_v^p + \delta_T u_v^n)/2 = (\delta_T u_v + \delta_T d_v)/2$, $(\delta_T d_v)_D = (\delta_T d_v + \delta_T u_v)/2$, for valence quark distributions, we obtain

$$\int dx\, b_1(x) = \frac{5}{36} \int dx \left[\delta_T u_v(x) + \delta_T d_v(x) \right] + \frac{1}{18} \int dx \left[8\, \delta_T \bar{u}(x) + 2\, \delta_T \bar{d}(x) + \delta_T s(x) + \delta_T \bar{s}(x) \right]_D, \tag{13}$$

[1] The overall factor 1/2 is introduced in b_1 as usual in defining F_1 and g_1 in terms of PDFs.

Journal of Physics: Conference Series **543** (2014) 012001 doi:10.1088/1742-6596/543/1/012001

where Q^2 dependence is abbreviated.

For describing the elastic scattering, we define the helicity amplitude with the charge operator J_0 by

$$\Gamma_{H,H} = \langle\, p,\ H \mid J_0(0) \mid p,\ H\, \rangle. \tag{14}$$

As shown in Fig. 3, the frame with large longitudinal momentum $|\vec{p}| \gg M$ is considered. Then, the parton model can be used for estimating the amplitude by assuming that the quarks move along the longitudinal direction with the momentum fraction x and we obtain

$$\Gamma_{H,H} = \sum_i e_i \int dx\, \left[q_i^H(x) - \bar{q}_i^H(x) \right]_D \equiv \sum_i e_i \int dx\, \left[q_{i,v}^H(x) \right]_D . \tag{15}$$

The tensor spin combination of the amplitudes is written in terms of the valence-quark distributions in the deuteron by the proton and neutron contributions

$$\Gamma_{0,0} - \frac{\Gamma_{1,1} + \Gamma_{-1,-1}}{2} = \frac{1}{3} \int dx\, \left[\delta_T u_v(x) + \delta_T d_v(x) \right], \tag{16}$$

where the relation $\int dx\, [\delta_T s(x) - \delta_T \bar{s}(x)] = 0$ is used. Substituting this relation into Eq. (13), we obtain

$$\int dx\, b_1(x) = \frac{5}{12} \left[\Gamma_{0,0} - \frac{\Gamma_{1,1} + \Gamma_{-1,-1}}{2} \right] + \frac{1}{9} \int dx\, \left[4\, \delta_T \bar{u}(x) + \delta_T \bar{d}(x) + \delta_T \bar{s}(x) \right]_D . \tag{17}$$

The next step is to calculate the tensor helicity combination by macroscopic quantities for the deuteron. The elastic scattering amplitudes can be described in terms of electric charge and quadrupole form factors, $F_C(t)$ and $F_Q(t)$ where t is the momentum-transfer squared, of the deuteron as

$$\Gamma_{0,0} = \lim_{t \to 0} \left[F_C(t) - \frac{t}{3M^2} F_Q(t) \right], \quad \Gamma_{1,1} = \Gamma_{-1,-1} = \lim_{t \to 0} \left[F_C(t) + \frac{t}{6M^2} F_Q(t) \right]. \tag{18}$$

Here, the $t \to 0$ limit is taken, and the units of F_C and F_Q are given by e and e/M^2, respectively, where e is the unit charge and M is the mass of the deuteron. Then, the tensor spin combination is expressed in terms of the quadrupole form factor as

$$\Gamma_{0,0} - \frac{\Gamma_{1,1} + \Gamma_{-1,-1}}{2} = - \lim_{t \to 0} \frac{t}{4} F_Q(t) = 0. \tag{19}$$

Substituting this relation into Eq. (17), we finally obtain

$$\int dx\, b_1(x) = - \lim_{t \to 0} \frac{5}{24} t F_Q(t) + \frac{1}{9} \int dx\, \left[4\, \delta_T \bar{u}(x) + \delta_T \bar{d}(x) + \delta_T \bar{s}(x) \right]_D . \tag{20}$$

We should note that the derived sum rule is not a rigorous one, but it is based on the parton model. The situation is the same as the Gottfried sum rule. There is a similarity between these sum rules:

$$\begin{aligned} \int dx\, b_1(x) &= 0 + \frac{1}{9} \int dx\, \left[4\, \delta_T \bar{u}(x) + \delta_T \bar{d}(x) + \delta_T \bar{s}(x) \right], \\ \int \frac{dx}{x} \left[F_2^p(x) - F_2^n(x) \right] &= \frac{1}{3} + \frac{2}{3} \int dx\, [\bar{u}(x) - \bar{d}(x)]. \end{aligned} \tag{21}$$

Journal of Physics: Conference Series **543** (2014) 012001 doi:10.1088/1742-6596/543/1/012001

The factor of 1/3 in the Gottfried sum rule comes from the flavor dependence of the valence quark distributions: $\int [u_v(x) - d_v(x)]/3 = 1/3$. In the same way, the factor $-\lim_{t\to 0}(5/24)tF_Q(t) = 0$ comes from the tensor-polarized valence quark distributions. The difference of zero and a finite number in both sum rules could be summarized as *the valence-quark number depends on flavor, whereas the number does not depend on the tensor polarization.* As the violation of the Gottfried sum rule led to fruitful studies of flavor asymmetric antiquark distributions $\bar{u} - \bar{d}$ and its physics origins [25], there is a good possibility that a finite value of the b_1 sum indicates the tensor-polarized antiquark distributions according to Eq. (21). There were HERMES measurements on b_1 and it indicated [17]

$$\int_{0.002}^{0.85} dx b_1(x) = [1.05 \pm 0.34 \text{ (stat)} \pm 0.35 \text{ (sys)}] \times 10^{-2},$$
$$\int_{0.02}^{0.85} dx b_1(x) = [0.35 \pm 0.10 \text{ (stat)} \pm 0.18 \text{ (sys)}] \times 10^{-2} \text{ for the range } Q^2 > 1\,\text{GeV}^2, \quad (22)$$

which suggests a finite tensor polarization for antiquarks. It is very interesting to find the physics origin of finite tensor polarized distributions. It will be tested by the approved JLab experiment on b_1 [18].

Both b_1 and Gottfried sum rules are obtained in the parton model, and they provide useful guidelines for the tensor-polarized antiquark distributions ($\delta_T \bar{q}$) and the light-quark flavor dependence of the unpolarized antiquark distributions ($\bar{u}/\bar{d}$). However, one should be careful that they are not rigorous sum rules. For example, the $\bar{u}(x)$ and $\bar{d}(x)$ distributions are assumed to be equal at very small x according to current parametrizations on the unpolarized PDFs. Therefore, the integral $\int dx(\bar{u} - \bar{d})$ seems to converge at this stage, but it could become infinite if a slight difference exists between $\bar{u}(x)$ and $\bar{d}(x)$ at small x. In the same way, the integral $\int \delta_T \bar{q}(x)$ may not be finite depending on the x dependence of $\delta_T \bar{q}(x)$ or on its Q^2 dependence. As noted in recent studies of Ref. [6], the sum rule are not satisfied in some theoretical models. In a convolution model of deuteron structure function, the b_1 sum is given by the product of a moment of lightcone momentum distribution of the nucleon with a moment of unpolarized quark distribution. The first nucleon part identically vanishes but the second PDF part could diverge at large Q^2, although it converges in some PDF parametrization at small Q^2, so that the whole integral is not certain. We need more theoretical efforts on an appropriate description of the tensor-polarized distributions, especially on the antiquark part, together with experimental measurements.

5. Parametrization of tensor-polarized distributions

Theoretical models are used for calculating structure functions, and there are studies on low moments of the structure functions by lattice QCD. However, it is almost impossible to obtain x-dependent distributions at this stage, so that global analyses of world experimental data are used for determining reliable PDFs. In the same way, a useful parametrization of the tensor polarized distributions can be proposed by using the existing HERMES data, although the number of data is not sufficient for an accurate determination. It should be useful for proposing future experiments (*e.g.* [18]) and for testing theoretical calculations.

We consider that certain fractions of the unpolarized distributions are tensor polarized:

$$\delta_T q_{iv}^D(x) = \delta_T w(x)\, q_{iv}^D(x), \quad \delta_T \bar{q}_i^D(x) = \alpha_{\bar{q}}\, \delta_T w(x)\, \bar{q}_i^D(x), \quad (23)$$

where $\delta_T w(x)$ and $\alpha_{\bar{q}}\, \delta_T w(x)$ are such fractions for valence quarks and antiquarks, respectively. The x dependence of $\delta_T w(x)$ could be different between valence quarks and antiquarks, and flavor dependence may exist. However, it is not the stage to investigate such details because

Journal of Physics: Conference Series **543** (2014) 012001 doi:10.1088/1742-6596/543/1/012001

Table 1. Determined parameters are listed for two sets of analyses.

Analysis	χ^2/d.o.f.	a	$\alpha_{\bar{q}}$	b	c	x_0
Set 1	2.83	0.378 ± 0.212	0.0 (fixed)	0.706 ± 0.324	1.0 (fixed)	0.229
Set 2	1.57	0.221 ± 0.174	3.20 ± 2.75	0.648 ± 0.342	1.0 (fixed)	0.221

the HERMES data are the only ones. The tensor-polarized distributions cannot be determined accurately at this stage, so that simplifying assumptions are employed for the unpolarized PDFs. Nuclear modifications in the PDFs of the deuteron are considered as a few percent effects [26], and they are neglected in our studies. Namely, the PDFs of the deuteron are given by the contributions from proton and neutron: $q_i^D = (q_i^p + q_i^n)/2$ and $\bar{q}_i^D = (\bar{q}_i^p + \bar{q}_i^n)/2$. Next, isospin symmetry is used for relating the PDFs of the neutron to the ones of the proton: $u_n = d$, $d_n = u$, $\bar{u}_n = \bar{d}$, and $\bar{d}_n = \bar{u}$, and flavor symmetric tensor-polarized distributions are assumed. Then, we have the distributions

$$\begin{aligned}\delta_T q_v^D(x) &\equiv \delta_T u_v^D(x) = \delta_T d_v^D(x) = \delta_T w(x)\,\frac{u_v(x)+d_v(x)}{2},\\ \delta_T \bar{q}^D(x) &\equiv \delta_T \bar{u}^D(x) = \delta_T \bar{d}^D(x) = \delta_T s^D(x) = \delta_T \bar{s}^D(x)\\ &= \alpha_{\bar{q}}\,\delta_T w(x)\,\frac{2\bar{u}(x)+2\bar{d}(x)+s(x)+\bar{s}(x)}{6},\end{aligned} \tag{24}$$

for an analysis of the HERMES data. Any available unpolarized PDFs could be used, but the LO version of the MSTW parametrization [27] is employed. From these tensor-polarized distributions, we finally obtain b_1 for the deuteron as

$$b_1^D(x) = \frac{1}{36}\delta_T w(x)\,\left[5\{u_v(x)+d_v(x)\} + 4\alpha_{\bar{q}}\{2\bar{u}(x)+2\bar{d}(x)+s(x)+\bar{s}(x)\}\right]. \tag{25}$$

The important point of the analysis is how to choose the x dependence of $\delta_T w(x)$. From the derivation of the b_1 sum rule, the tensor-polarized valence-quark distributions should satisfy the sum $\int dx (b_1)_{\text{valence}} = 0$. In order to satisfy this relation, the function $\delta_T w(x)$ should have a node at least, so we may take the parametrization

$$\delta_T w(x) = a x^b (1-x)^c (x_0 - x), \tag{26}$$

where a, b, c, and x_0 are the parameters determined by the analysis. The existence of the node is also supported by the convolution model for b_1 with a D-state admixture. However, if the first moment of $\delta_T w(x)$ vanishes, x_0 can be expressed by other ones as

$$x_0 = \frac{\int_0^1 dx x^{b+1}(1-x)^c\{u_v(x)+d_v(x)\}}{\int_0^1 dx x^b (1-x)^c\{u_v(x)+d_v(x)\}}. \tag{27}$$

There is no data to probe the scaling violation at this stage, so that the Q^2 dependence is neglected in this analysis. For calculating the unpolarized PDFs Eq. (25), the average Q^2 value (Q^2=2.5 GeV^2) of the HERMES experiment is used.

In order to find the impact of the tensor-polarized antiquark distributions, we made two types of analyses:

- Set 1: Tensor-polarized antiquark distributions are terminated ($\alpha_{\bar{q}} = 0$).
- Set 2: Finite tensor-polarized antiquark distributions are allowed ($\alpha_{\bar{q}}$ is a parameter).

Journal of Physics: Conference Series **543** (2014) 012001 doi:10.1088/1742-6596/543/1/012001

The constant c is fixed at $c = 1$ in these analyses due to the lack of data at large x to constrain it. The obtained parameters are listed in Table 1. The smaller χ^2 value of the set-2 analysis indicates that there is a significant improvement in the fit by including the tensor-polarized antiquark distributions.

Obtained b_1 structure functions are compared with the HERMES data in Fig. 4. The dashed and solid curves are for the set-1 and the set-2, respectively. A reasonable fit was obtained by the set-2 analysis, whereas the small-x (< 0.1) data cannot be explained by the set-1. Namely, the HERMES data at small x suggest the existence of a finite tensor-polarized antiquark distributions according to this analysis. The medium and large x regions are well explained by the tensor-polarized valence-quark distributions.

Determined tensor polarized distributions are shown in Fig. 5. The dashed and solid curves are tensor-polarized valence-quark distributions for the set-1 and set-2, respectively, and the dotted curve indicates the tensor-polarized antiquark distribution of the set-2. The valence-quark distribution is negative at medium x and it turned to negative at $x < 0.2$. The antiquark distribution becomes large at $x < 0.2$. The functional form of the valence-quark distribution is expected in the convolution description of the deuteron structure function [5]. However, the physics origin of the antiquark distribution is not obvious at this stage. More theoretical efforts are needed to understand the origin. From the obtained distributions of the set-2, the b_1 sum is estimated as

$$\int dx\, b_1(x) = -\frac{5}{24}\lim_{t\to 0} tF_Q(t) + \frac{1}{18}\int dx\,[\,8\delta_T\bar{u}(x) + 2\delta_T\bar{d}(x) + \delta_T s(x) + \delta_T\bar{s}(x)\,]$$
$$= 0.0058. \qquad (28)$$

This finite value is due to the existence of antiquark tensor polarization. There will be measurements on b_1 at JLab [18], so that more details will become clear experimentally. On the other hand, theoretical studies are needed for the tensor-polarized quark and antiquark distributions.

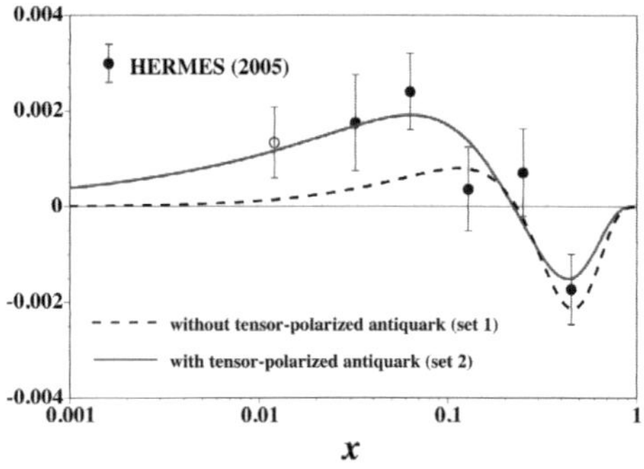

Figure 4. Obtained b_1 structure functions are compared with HERMES data.

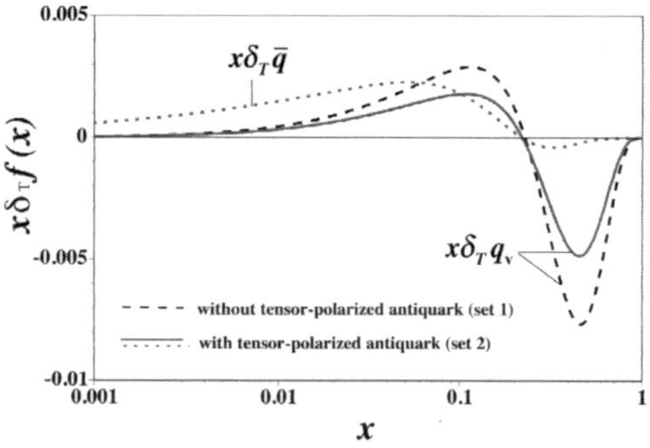

Figure 5. Determined tensor polarized distributions are shown.

6. Polarized proton-deuteron Drell-Yan processes

6.1. Structure functions and spin asymmetries

The HERMES data indicated an existence of finite tensor-polarized antiquark distributions. One of methods to probe antiquark distributions is to use Drell-Yan processes. For example, the polarized proton-deuteron (pd) Drell-Yan processes could be used with tensor-polarized deuterons. There is no experiment for the pd Drell-Yan, and there are only a few theoretical formalisms [16].

We consider the Drell-Yan process of $A\,(\text{spin}\,1/2) + B\,(\text{spin}\,1) \to \ell^+\ell^- + X$. Our formalism can be used for any spin-1/2 and spin-1 hadrons, but the most realistic reaction is the proton-deuteron Drell-Yan. Its cross section is written in terms of the lepton tensor $L_{\mu\nu}$ and the hadron tensor $W^{\mu\nu}$

$$\frac{d\sigma}{d^4Q\,d\Omega} = \frac{\alpha^2}{2\,s\,Q^4}\,L_{\mu\nu}\,W^{\mu\nu}\,, \tag{29}$$

where $\alpha = e^2/(4\pi)$ is the fine structure constant, s is the center-of-mass energy squared $s = (P_A + P_B)^2$, Q is the total dilepton momentum, and Ω is the solid angle of the momentum $\vec{k}_{\ell+} - \vec{k}_{\ell-}$.

A general spin-density formalism, by expressing structure functions in terms of helicity amplitudes and Clebash-Gordan coefficients with the conditions of Hermiticity, parity conservation, and time-reversal invariance, indicates that there exist 108 structure functions for the unpolarized and polarized pd Drell-Yan processes, whereas there are 48 structure functions for the pp. There are 60 new structure functions, which should be associated with the deuteron tensor structure. Of course, all of them are not important for the first investigation. If the cross section is integrated over the lepton-pair transverse momentum $\vec{Q}_T$, there are only 22 ones which include 11 new functions associated with the tensor structure of the deuteron. This spin-density formalism is rather lengthy, so that we refer the paper [16] for the details.

In the general hadron tensor formalism, the hadron tensor of the Drell-Yan process

$$W^{\mu\nu} = \int \frac{d^4\xi}{(2\,\pi)^4}\, e^{iQ\cdot\xi}\,\langle\, P_A S_A P_B S_B\,|\, J^\mu(0) J^\nu(\xi)\,|\, P_A S_A P_B S_B\,\rangle, \tag{30}$$

is expanded in terms of possible Lorentz index combinations including the hadron momenta and spins by considering the conditions

$$\begin{aligned}
&\text{Hermiticity:} && [W^{\nu\mu}(Q; P_A S_A; P_B S_B)]^* = W^{\mu\nu}(Q; P_A S_A; P_B S_B),\\
&\text{Parity conservation:} && W^{\mu\nu}(Q; P_A S_A; P_B S_B) = W_{\mu\nu}(\overline{Q}; \overline{P}_A - \overline{S}_A; \overline{P}_B - \overline{S}_B),\\
&\text{Time-reveral invariance:} && [W^{\mu\nu}(Q; P_A S_A; P_B S_B)]^* = W_{\mu\nu}(\overline{Q}; \overline{P}_A \overline{S}_A; \overline{P}_B \overline{S}_B),
\end{aligned} \tag{31}$$

where $\overline{P}$ is defined by $\overline{P}^\mu = (P^0, -\vec{P})$. For the expansion, the Lorentz vectors X^μ, Y^μ, and Z^μ

$$\begin{aligned}
&X^\mu = P_A^\mu\, Q^2\, Z \cdot P_B - P_B^\mu\, Q^2\, Z \cdot P_A + Q^\mu\,(Q \cdot P_B\, Z \cdot P_A - Q \cdot P_A\, Z \cdot P_B),\\
&Y^\mu = \epsilon^{\mu\alpha\beta\gamma}\, P_{A\alpha}\, P_{B\beta}\, Q_\gamma, \quad Z^\mu = P_A^\mu\, Q \cdot P_B - P_B^\mu\, Q \cdot P_A,
\end{aligned} \tag{32}$$

are used. We also define the vector T^μ by

$$T^\mu = \varepsilon^{\mu\alpha\beta\gamma} S_\alpha\, Z_\beta\, Q_\gamma\,. \tag{33}$$

In addition to these vectors, Q^μ, S_A^μ, S_B^μ, T_A^μ, and T_B^μ can be used to expand $W^{\mu\nu}$. However, instead of S_A^μ and S_B^μ, it is more convenient to use the transverse vectors S_{AT}^μ and S_{BT}^μ:

$$S_T^\mu = \left(g^{\mu\nu} - \frac{Q^\mu Q^\nu}{Q^2} - \frac{Z^\mu Z^\nu}{Z^2} \right) S_\nu\,. \tag{34}$$

In the general case, it is too lengthy to write them down here becasuse there are 108 structure functions. The transverse momentum Q_T is roughly restricted by the hadron size R by

$Q_T < 1/R$, so that the limit of $Q_T \to 0$ is considered in the following formalism. Then, X^μ and Y^μ do not have to be considered because $|\vec{X}|$ and $|\vec{Y}|$ are proportional to Q_T and $X^0 = Y^0 = 0$ in the dilepton rest frame. Then, we obtain

$$\begin{aligned} W^{\mu\nu} = & - g^{\mu\nu} A - \frac{Z^\mu Z^\nu}{Z^2} B + Z^{\{\mu} T_A^{\nu\}} C + Z^{\{\mu} T_B^{\nu\}} D + Z^{\{\mu} S_{AT}^{\nu\}} E + Z^{\{\mu} S_{BT}^{\nu\}} F \\ & - S_{BT}^\mu S_{BT}^\nu G - S_{AT}^{\{\mu} S_{BT}^{\nu\}} H + T_A^{\{\mu} S_{BT}^{\nu\}} I + S_{BT}^{\{\mu} T_B^{\nu\}} J + Q^\mu Q^\nu K + Q^{\{\mu} Z^{\nu\}} L \\ & + Q^{\{\mu} S_{AT}^{\nu\}} M + Q^{\{\mu} S_{BT}^{\nu\}} N + Q^{\{\mu} T_A^{\nu\}} O + Q^{\{\mu} T_B^{\nu\}} P , \end{aligned} \tag{35}$$

where $Q^{\{\mu} Z^{\nu\}}$ is defined by $Q^{\{\mu} Z^{\nu\}} \equiv Q^\mu Z^\nu + Q^\nu Z^\mu$. Next, we impose the current conservation $Q_\mu W^{\mu\nu} = 0$, and then the coefficients A, B, $\cdots$, which still contain spin factors, are expanded by the scalar and pseudoscalar terms with the spins. We finally obtain

$$\begin{aligned} W^{\mu\nu} = & - \left[g^{\mu\nu} - \frac{Q^\mu Q^\nu}{Q^2} \right] \bigg\{ W_{0,0} + \frac{M_A M_B}{s\, Z^2} Z \cdot S_A \, Z \cdot S_B \, V_{0,0}^{LL} - S_{AT} \cdot S_{BT} \, V_{0,0}^{TT} \\ & \qquad - \left(\frac{8\, M_B^2 \, (Z \cdot S_B)^2}{s^2 \, (Q \cdot P_B)^2} + \frac{4}{3} S_B^2 \right) V_{0,0}^{UQ_0} + \frac{M_B}{Z^2 \, Q \cdot P_B} Z \cdot S_B \, T_A \cdot S_{BT} \, V_{0,0}^{TQ_1} \bigg\} \\ & - \left[\frac{Z^\mu Z^\nu}{Z^2} - \frac{1}{3} \left(g^{\mu\nu} - \frac{Q^\mu Q^\nu}{Q^2} \right) \right] \bigg\{ W_{2,0} + \frac{M_A M_B}{s\, Z^2} Z \cdot S_A \, Z \cdot S_B \, V_{2,0}^{LL} - S_{AT} \cdot S_{BT} \, V_{2,0}^{TT} \\ & \qquad - \left(\frac{8\, M_B^2 \, (Z \cdot S_B)^2}{s^2 \, (Q \cdot P_B)^2} + \frac{4}{3} S_B^2 \right) V_{2,0}^{UQ_0} + \frac{M_B}{Z^2 \, Q \cdot P_B} Z \cdot S_B \, T_A \cdot S_{BT} \, V_{2,0}^{TQ_1} \bigg\} \\ & - Z^{\{\mu} T_A^{\nu\}} \frac{1}{\sqrt{Q^2} \, Z^2} \bigg\{ U_{2,1}^{TU} - \left(\frac{8\, M_B^2 \, (Z \cdot S_B)^2}{s^2 \, (Q \cdot P_B)^2} + \frac{4}{3} S_B^2 \right) U_{2,1}^{TQ_0} \bigg\} \\ & - Z^{\{\mu} T_B^{\nu\}} \frac{1}{\sqrt{Q^2} \, Z^2} \bigg\{ U_{2,1}^{UT} + \frac{M_A M_B}{s\, Z^2} Z \cdot S_A \, Z \cdot S_B \, U_{2,1}^{LQ_1} + S_{AT} \cdot S_{BT} \, U_{2,1}^{TQ_2} \bigg\} \\ & + Z^{\{\mu} S_{AT}^{\nu\}} \frac{\sqrt{Q^2} \, M_B}{Z^2 \, Q \cdot P_B} Z \cdot S_B \, U_{2,1}^{TL} \\ & + Z^{\{\mu} S_{BT}^{\nu\}} \bigg\{ - \frac{\sqrt{Q^2} \, M_A}{Z^2 \, Q \cdot P_A} Z \cdot S_A \, U_{2,1}^{LT} + \frac{\sqrt{Q^2} \, M_B}{Z^2 \, Q \cdot P_B} Z \cdot S_B \, U_{2,1}^{UQ_1} - \frac{1}{\sqrt{Q^2} \, Z^2} T_A \cdot S_{BT} \, U_{2,1}^{TQ_2} \bigg\} \\ & - \left[2\, S_{BT}^\mu S_{BT}^\nu - S_{BT}^2 \left(g^{\mu\nu} - \frac{Q^\mu Q^\nu}{Q^2} - \frac{Z^\mu Z^\nu}{Z^2} \right) \right] U_{2,2}^{UQ_2} \\ & - \left[S_{AT}^{\{\mu} S_{BT}^{\nu\}} - S_{AT} \cdot S_{BT} \left(g^{\mu\nu} - \frac{Q^\mu Q^\nu}{Q^2} - \frac{Z^\mu Z^\nu}{Z^2} \right) \right] U_{2,2}^{TT} \\ & - \left[T_A^{\{\mu} S_{BT}^{\nu\}} - T_A \cdot S_{BT} \left(g^{\mu\nu} - \frac{Q^\mu Q^\nu}{Q^2} - \frac{Z^\mu Z^\nu}{Z^2} \right) \right] \frac{M_B}{Z^2 \, Q \cdot P_B} Z \cdot S_B \, U_{2,2}^{TQ_1} \\ & + S_{BT}^{\{\mu} T_B^{\nu\}} \frac{M_A}{Z^2 \, Q \cdot P_A} Z \cdot S_A \, U_{2,2}^{LQ_2} \, . \end{aligned} \tag{36}$$

Therefore, in the limit of $Q_T \to 0$, threre are 22 structure functions:

$$\begin{aligned} & W_{0,0}, \; V_{0,0}^{LL}, \; V_{0,0}^{TT}, \; V_{0,0}^{UQ_0}, \; V_{0,0}^{TQ_1}, W_{2,0}, \; V_{2,0}^{LL}, \; V_{2,0}^{TT}, \; V_{2,0}^{UQ_0}, V_{2,0}^{TQ_1}, \; U_{2,1}^{TU}, \\ & U_{2,1}^{TQ_0}, \; U_{2,1}^{UT}, \; U_{2,1}^{LQ_1}, \; U_{2,1}^{TQ_2}, \; U_{2,1}^{TL}, \; U_{2,1}^{LT}, \; U_{2,1}^{UQ_1}, \; U_{2,2}^{UQ_2}, \; U_{2,2}^{TT}, \; U_{2,2}^{TQ_1}, \; U_{2,2}^{LQ_2}, \end{aligned} \tag{37}$$

where W, V, and U are an unpolarized structure function, a polarized one without the spin factors in the hadron tensor, and a polarized one with the spin factor. The function $W_{L,M}$

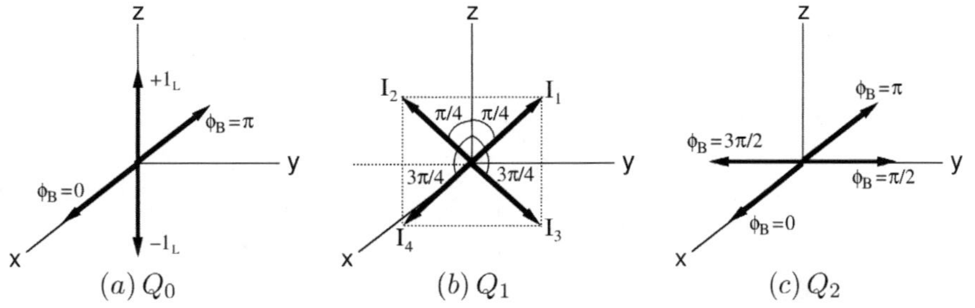

Figure 6. Tensor polarizations Q_0, Q_1, and Q_2.

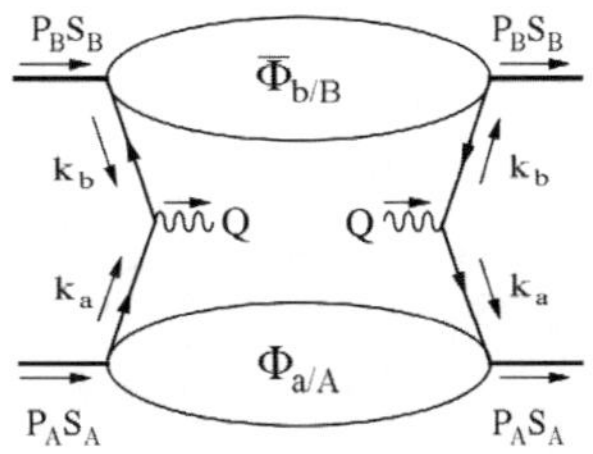

Figure 7. Parton model for Drell-Yan.

is obtained by the integral $\int d\Omega\, Y_{LM}\, d\sigma/(d^4Q\, d\Omega) \propto W_{L,M}$ of the unpolarized reaction. The superscripts U, L, and T show unpolarized, longitudinally polarized, and transversely polarized states. The quadrupole polarizations Q_0, Q_1, and Q_2 are associated with the spherical harmonics Y_{20}, Y_{21}, and Y_{22} as shown in Fig. 6. They are the polarizations in the xz, yz, and xy planes. The structure functions with Q_0, Q_1, and Q_2 are specific for the spin-1 deuteron.

In the pp Drell-Yan processes, the unpolarized, longitudinal, and transverse combinations exist: $<\sigma>$, A_{LL}, A_{TT}, A_{LT}, and A_T. In addition, the following fifteen quadrupole spin asymmetries could be investigated in the pd Drell-Yan:

$$\begin{aligned}&<\sigma>,\ A_{LL},\ A_{TT},\ A_{LT},\ A_{TL},\ A_{UT},\ A_{TU},\\ &A_{UQ_0},\ A_{TQ_0},\ A_{UQ_1},\ A_{LQ_1},\ A_{TQ_1},\ A_{UQ_2},\ A_{LQ_2},\ A_{TQ_2}.\end{aligned} \tag{38}$$

These asymmetries are expressed in terms of the structure functions in Eq. (37). For example, the quadrupole spin asymmetry A_{UQ_0} is measured with the unpolarized proton and the Q_0-type tensor polarized deuteron, and it is expressed in terms of the structure functions $V_{0,0}^{UQ_0}$, $V_{2,0}^{UQ_0}$, $W_{0,0}$, and $W_{2,0}$:

$$A_{UQ_0} = \frac{1}{2<\sigma>}\left[\sigma(\bullet, 0_L) - \frac{\sigma(\bullet,+1_L)+\sigma(\bullet,-1_L)}{2}\right] = \frac{2\,V_{0,0}^{UQ_0} + (\frac{1}{3} - cos^2\theta)\,V_{2,0}^{UQ_0}}{2\,W_{0,0} + (\frac{1}{3} - cos^2\theta)\,W_{2,0}}, \tag{39}$$

where $\bullet$ indicates the unpolarized case.

6.2. Parton model expressions

Possible structure functions and spin asymmetries were introduced for the pd Drell-Yan processes. Here, we express them in terms of parton distribution functions of the proton and deuteron [16]. As shown in Fig. 7. The hadron tensor $W^{\mu\nu}$ of the Drell-Yan processes is

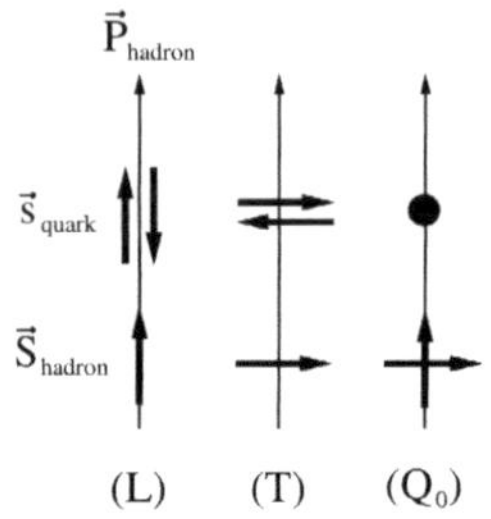

Figure 8. Longitudinally polarized, transversity, and tensor polarized distributions. The notation • indicates unpolarized.

written by the quark and antiquark correlation functions, which are expressed by combinations of possible vectors and pseudovectors. The leading contribution to the hadron tensor is

$$W^{\mu\nu} = \frac{1}{3}\sum_{a,b} \delta_{b\bar{a}}\, e_a^2 \int d^4k_a\, d^4k_b\, \delta^4(k_a + k_b - Q)\, Tr[\Phi_{a/A}(P_A S_A; k_a)\gamma^\mu \bar{\Phi}_{b/B}(P_B S_B; k_b)\gamma^\nu]\,. \quad (40)$$

The correlation functions $\Phi_{a/A}$ and $\bar{\Phi}_{\bar{a}/B}$ are defined by

$$\Phi_{a/A}(P_A S_A; k_a)_{ij} = \int \frac{d^4\xi}{(2\pi)^4}\, e^{ik_a\cdot\xi}\, \langle\, P_A S_A \,|\, \bar{\psi}_j^{(a)}(0)\, \psi_i^{(a)}(\xi) \,|\, P_A S_A \,\rangle\,,$$
$$\bar{\Phi}_{\bar{a}/B}(P_B S_B; k_{\bar{a}})_{ij} = \int \frac{d^4\xi}{(2\pi)^4}\, e^{ik_{\bar{a}}\cdot\xi}\, \langle\, P_B S_B \,|\, \psi_i^{(a)}(0)\, \bar{\psi}_j^{(a)}(\xi) \,|\, P_B S_B \,\rangle\,, \quad (41)$$

where link operators for the gauge invariance are not explicitly written. Using a Fierz transformation, we write the hadron tensor in a factorized form:

$$\begin{aligned} W^{\mu\nu} =& \frac{1}{3}\sum_{a,b} \delta_{b\bar{a}}\, e_a^2 \int d^2\vec{k}_{aT}\, d^2\vec{k}_{bT}\, \delta^2(\vec{k}_{aT} + \vec{k}_{bT} - \vec{Q}_T) \Bigg[\Bigg\{ - \Phi_{a/A}[\gamma^\alpha]\, \bar{\Phi}_{b/B}[\gamma_\alpha] \\ & - \Phi_{a/A}[\gamma^\alpha\gamma_5]\, \bar{\Phi}_{b/B}[\gamma_\alpha\gamma_5] + \frac{1}{2}\Phi_{a/A}[i\sigma_{\alpha\beta}\gamma_5]\, \bar{\Phi}_{b/B}[i\sigma^{\alpha\beta}\gamma_5] \Bigg\} g^{\mu\nu} + \Phi_{a/A}[\gamma^{\{\mu}]\, \bar{\Phi}_{b/B}[\gamma^{\nu\}}] \\ & + \Phi_{a/A}[\gamma^{\{\mu}\gamma_5]\, \bar{\Phi}_{b/B}[\gamma^{\nu\}}\gamma_5] + \Phi_{a/A}[i\sigma^{\alpha\{\mu}\gamma_5]\, \bar{\Phi}_{b/B}[i\sigma^{\nu\}}{}_\alpha\gamma_5] \Bigg] + O(1/Q)\,. \end{aligned} \quad (42)$$

Then, these correlation functions are expressed by the unpolarized, longitudinally-polarized, and transversity distributions of the proton and deuteron, together with the tensor-polarized distributions of the deuteron, as illustrated in Fig. 8. Particularly, it is important that the correlation function $\Phi[\gamma^\mu]$ contains the tensor-polarized distributions $\delta_T q(x)$. The details are found in Ref. [16].

In the naive parton model, we find 19 structure functions, which become four by the $\vec{Q}_T$ integration. We define $\overline{W} = \int d^2\vec{Q}_T\, W$, and $\overline{V}$ and $\overline{U}$ are defined in the same way. The pd Drell-Yan cross section is then given by

$$\begin{aligned} \frac{d\sigma}{dx_A\, dx_B\, d\Omega} =& \frac{\alpha^2}{4\,Q^2} \Bigg[(1+\cos^2\theta) \Bigg\{ \overline{W}_T + \frac{1}{4}\lambda_A\lambda_B\, \overline{V}_T^{LL} + \frac{2}{3}\left(2\,|\vec{S}_{BT}|^2 - \lambda_B^2\right) \overline{V}_T^{UQ_0} \Bigg\} \\ & + \sin^2\theta\, |\vec{S}_{AT}|\, |\vec{S}_{BT}| \cos(2\phi - \phi_A - \phi_B)\, \overline{U}_{2,2}^{TT} \Bigg]\,. \end{aligned} \quad (43)$$

The structure functions are expressed by the parton distributions in the process q(in p)+$\bar{q}$(in d)$\to \ell^+ + \ell^-$ as

$$\overline{W}_T = \frac{1}{3}\sum_i e_i^2\, q_i(x_A)\,\bar{q}_i(x_B)\,, \qquad \overline{V}_T^{LL} = -\frac{4}{3}\sum_i e_i^2\, \Delta q_i(x_A)\,\Delta\bar{q}_i(x_B)\,,$$

$$\overline{U}_{2,2}^{TT} = \frac{1}{3}\sum_i e_i^2\, \Delta_T q_i(x_A)\,\Delta_T\bar{q}_i(x_B)\,, \qquad \overline{V}_T^{UQ_0} = \frac{1}{6}\sum_i e_i^2\, q_i(x_A)\,\delta_T\bar{q}_i(x_B)\,, \tag{44}$$

where Δq_i and $\Delta_T q_i$ are longitudinally-polarized and transversity distributions. The tensor-polarized distributions can be studied by the asymmetry Q_0:

$$A_{UQ_0} = \frac{\overline{V}_T^{UQ_0}}{\overline{W}_T} = \frac{\sum_a e_a^2\,[\,q_a(x_A)\,\delta_T\bar{q}_a(x_B) + \bar{q}_a(x_A)\,\delta_T q_a(x_B)\,]}{2\sum_a e_a^2\,[\,q_a(x_A)\,\bar{q}_a(x_B) + \bar{q}_a(x_A)\,q_a(x_B)\,]}\,. \tag{45}$$

This asymmetry indicates that the tensor-polarized distributions $\delta_T q$ and $\delta_T \bar{q}$ should be found in the Drell-Yan in addition to the charged-lepton scattering. It is especially important that the antiqurk distribution $\delta_T \bar{q}$ is measured, because it was suggested that the finite $\delta_T \bar{q}$ was indicated by the HERMES experiment.

The possibility of polarized deuteron acceleration was once considered at RHIC [28], but it was not attained. However, there are future possibilities to investigate the Drell-Yan process with a fixed tensor-polarized deuteron target at hadron faclities such as Fermilab [21], J-PARC [22], GSI [23], and CERN-COMPASS. The JLab b_1 measurement will clarify the details of the tensor-polarized distributions in 2020's. Together with their data, we expect that the Drell-Yan measurements will clarify the tensor-polarized antiquark distributions.

7. Summary

We explained our studies on the tensor-polarized structure function b_1 and tensor-polarized quark and antiquark distributions, $\delta_T q$ and $\delta_T \bar{q}$. First, the projection operators are shown for all the eight structure functions of the deuteron from the hadron tensor $W^{\mu\nu}$. The projection operators should be useful in a convolution description of the deuteron structure functions. Second, the sum rule was explained for b_1 by using the parton model. It is valuable for indicating the existence of antiquark tensor polarization, as the Gottfried sum rule violation indicated a $\bar{u}/\bar{d}$ asymmetry in the nucleon. Third, the parametrization of the tensor-polarized quark and antiquark distributions was proposed by analyzing the HERMES data on b_1. The analysis indicated an existence of antiquark tensor polarization, and its origin should be investigated theoretically. The tensor-polarized antiquark distributions should be studied by the Drell-Yan processes with tensor-polarized deuteron. We showed the general formalisms for the structure functions in the proton-deuteron Drell-Yan processes and also their expressions in terms of the parton distribution functions in the proton and deuteron. With the tensor-polarized deuteron, the Drell-Yan process probes the tensor-polarized antiquark distributions directly. The JLab experiment on b_1 was approved and the actual measurement is expected to start in 2019. There are other possibilities to investigate the tensor structure at EIC, Fermilab, J-PARC, GSI, and CERN-COMPASS. The studies of tensor-polarized structure functions could open a new era of high-energy spin physics.

References

[1] Frankfurt L L and Strikman M I 1983 *Nucl. Phys. A* **405** 557
[2] Hoodbhoy P, Jaffe R L, and Manohar A, 1989 *Nucl. Phys. B* **312** 571; Jaffe R L and Manohar A 1989 *Nucl. Phys. B* **321** 343

[3] Close F E and Kumano S 1990 *Phys. Rev. D* **42** 2377
[4] Bacchetta A and Mulders P J 2000 *Phys. Rev. D* **62** 114004
[5] Khan H and Hoodbhoy P, 1991 *Phys. Rev. C* **44** 1219
[6] Miller G A 1990 pp.30-33 in *Topical Conference on Electronuclear physics with Internal Targets*, edited by R. G. Arnold (World Scientific, Singapore); Miller G A, 2014 *Phys. Rev. D* **89** 045203
[7] Nikolaev N N and Schäfer W 1997 *Phys. Lett. B* **398** 245; Nikolaev N N and Schäfer W 1997 *Phys. Lett. B* **407**, 453; Edelmann J, Piller G, and Weise W 1997 *Z. Phys. A* **357** 129
[8] Bora K and Jaffe R L 1998 *Phys. Rev. D* **57** 6906
[9] Schäfer A, Szymanowski L, and Teryaev O V 1999 *Phys. Lett. B* **464** 94
[10] Berger E R, Cano F, Diehl M, and Pire B 2001 *Phys. Rev. Lett.* **87** 142302; Kirchner and D. Mueller 2003 *Eur. Phys. J. C* **32** 347; Diehl M 2003 *Phys. Rept.* **388** 41; Cano F and Pire B 2004 *Eur. Phys. J. A* **19** 423; Belitsky A V and Radyushkin A V 2005 *Phys. Rept.* **418** 1
[11] Detmold W 2006 *Phys. Lett. B* **632** 261
[12] Dmitrasinovic V 1996 *Phys. Rev. D* **54** 1237
[13] Best C *et al.* 1997 *Phys. Rev. D* **56** 2743
[14] Kimura T-Y and Kumano S 2008 *Phys. Rev. D* **78** 117505
[15] Taneja S K, Kathuria K, Liuti S, and Goldstein G R 2012 *Phys. Rev. D* **86** 036008
[16] Hino S and Kumano S 1999 *Phys. Rev. D* **59** 094026; Hino S and Kumano S 1999 *Phys. Rev. D* **60** 054018; Kumano S and Miyama M 2000 *Phys. Lett. B* **479** 149
[17] Airapetian A *et al* 2005 *Phys. Rev. Lett.* **95** 242001; Riedl C 2014 Talk at the Tensor Polarized Solid Target Workshop http://www.jlab.org/conferences/tensor2014/
[18] Slifer K *et al* 2012 E12-13-011 Proposal to Jefferson Lab PAC-40; Slifer K 2014 Talk at the Tensor Polarized Solid Target Workshop http://www.jlab.org/conferences/tensor2014/
[19] Weiss C and Kalantarians N 2014 Talks at the Tensor Polarized Solid Target Workshop http://www.jlab.org/conferences/tensor2014/
[20] Long E, Strikman M, Sargsian M, and Cosyn W 2014 Talks at the Tensor Polarized Solid Target Workshop http://www.jlab.org/conferences/tensor2014/; Long E *et al* 2014 LOI12-14-002 Letter of Intent to Jefferson Lab PAC42
[21] Jiang X, 2014 Private communications on Fermilab Drell-Yan experiment
[22] Kumano S 2007 *Nucl. Phys. A* **782** 442; Kumano S 2008 *AIP Conf. Proc.* **1056** 444
[23] http://www.gsi.de/fair/index_e.html.
[24] R. P. Feynman 1998 *Photon-Hadron Interactions* (Westview press)
[25] Kumano S 1998 *Phys. Rept.* **303** 183; Garvey G T and Peng J-C 2001 *Prog. Part. Nucl. Phys.* **47** 203; Peng J-C and Qiu J-W 2014 *Prog. Part. Nucl. Phys.* **76** 43
[26] Hirai M, Kumano S, and Nagai T-H 2007 *Phys. Rev. C* **76** 065207
[27] Martin A D, Stirling W J, Thorne R S, and Watt G 2009 *Eur. Phys. J. C* **63** 189
[28] Courant E D 1998 Report BNL-65606

Journal of Physics: Conference Series **543** (2014) 012002 doi:10.1088/1742-6596/543/1/012002

Pionic and hidden-color, six-quark contributions to the deuteron b_1 structure function

G A Miller

Univ. of Washington, Seattle, WA 98195, USA

E-mail: `miller@uw.edu`

Abstract. The b_1 deep-inelastic structure function is an observable feature of a spin-1 system that is sensitive to non-nucleonic components of the target nuclear wave function. The contributions of exchanged pions in the deuteron are estimated and found to be of measurable size for values of x of about 0.1. A simple model for a hidden-color, six-quark configuration (with only about 0.15% probability to exist in the deuteron) is proposed and found to give substantial contributions for values of x greater than about 0.2. Good agreement with the only existing (HERMES) data is obtained. Predictions are made for an upcoming JLab experiment. The Close and Kumano sum rule is investigated and found to be a useful guide to understanding various possible effects that may contribute.

1. Introduction

This conference proceeding is based entirely on previously published work [1].

Consider lepton-nucleus deep-inelastic scattering, with the direction of the photon as the spin quantization axis. The cross section for an unpolarized lepton but polarized deuteron (D) target depends on the value m of the magnetic quantum number of the target:

$$d^2\sigma^{(m)} \propto l^{\mu\nu} W_{\mu\nu}^{(m)}, \tag{1}$$

where $l^{\mu\nu}$ is the standard lepton tensor and

$$W_{\mu\nu}^{(m)} = \int d^4 r e^{iq\cdot r} \langle D, J=1, J_z=m | [j^\mu(r), j^\nu(0)] | D, J=1, J_z=m \rangle. \tag{2}$$

The standard structure function F_1 is given by

$$F_1(x) = \frac{1}{3}\sum_m W_{11}^{(m)}, \tag{3}$$

and we are interested in

$$b_1(x) = W_{11}^{(1)} - W_{11}^{(0)}, \tag{4}$$

with x as the Bjorken variable, and we ignore the Q^2 dependence for simplicity. The function b_1 has been measured by the HERMES collaboration collaboration using a tensor polarized deuteron target [2] for values of Bjorken $0.01 < x < 0.45$. The function b_1 takes on its largest value of about 10^{-2} at the lowest measured value of x (0.012), decreases with increasing x through zero and takes on a minimum value of roughly -4×10^{-3}.

The function b_1 nearly vanishes if the spin-one target is made of constituents in a relative s-state, and is very small for a target of spin 1/2 particles moving non-relativistically in higher angular momentum states [3, 4]. Thus one expects [3] that a nuclear b_1 may be dominated by non-nucleonic components of the target nuclear wave function. Consequently, a Jefferson Laboratory experiment [5] is planned to measure b_1 for values of x in the range $0.16 < x < 0.49$ and $1 < Q^2 < 5$ GeV^2 with the aim of reducing the error bars.

2. General Remarks

The quark distribution function (related to F_1) can be written as

$$q^{(m)}(x) = \langle D, J = 1, m|O|D, J = 1, m\rangle. \tag{5}$$

If $q^{(m)}(x)$ is to have any dependence on m, the operator O can not be scalar. The Wigner Eckart theorem then dictates the O is a tensor of rank 1 or 2, if b_1 is to be different from zero. But invariance under parity says that $q^{(1)} = q^{(-1)}$. Thus O can not be of rank 1 and give an non-zero b_1.

This means that O is a rank 2 tensor, so that b_1 measures tensor effects. A consequence is that the S wave component of the deuteron gives no contribution to b_1 because the value of m is not relevant. In general, the nucleon contributions do not give anything large enough to be observable [3, 4].

3. Pionic Contribution

This contribution is detailed in [1], which displays the relevant figures. We summarize here. The pionic contribution to b_1 is given by

$$b_1^\pi(x) = \frac{1}{2}\int_x^2 \frac{dy}{y} q^\pi(x/y)\delta f_\pi(y), \tag{6}$$

where

$$\delta f_\pi(y) \equiv f_\pi^{(0)}(y) - f_\pi^{(1)}(y) \tag{7}$$

with

$$f_\pi^{(m)}(y_A) = \int \frac{d\xi^-}{2\pi} e^{-iy_A P_D^+ \xi^-} \langle D, m|\phi_\pi(\xi^-)\phi_\pi(0)|D, m\rangle_c \tag{8}$$

gives the probability for the pion to have a light cone momentum fraction of y. The function $q^\pi(x/y)$ is the quark distribution function of the pion.

The function $\delta f_\pi(y)$ has some interesting properties. It is independent of the deuteron wave function and has a double node structure which is a consequence of the tensor nature of b_1. Indeed, we find

$$\int_0^2 dy \frac{f_{abL}(y)}{y} = 0. \tag{9}$$

The pionic effects on b_1 are found to be substantial for x less than about 0.2, and for those values of x are large enough to account for the HERMES data. See Table I and Fig. 3 of Ref. [1].

Table 1. Measured values (in 10^{-2} units) of the tensor structure function b_1. Both the statistical and systematic uncertainties are listed. The numbers in parenthesis refer to the structure function modes of Ref. [6] .

$\langle x \rangle$	$\langle Q^2 \rangle$ [GeV^2]	b_1 [10^{-2}]	$\pm\delta b_1{}^{stat}$ [10^{-2}]	$\pm\delta b_1{}^{sys}$ [10^{-2}]	$b_1^{\pi}[7]$ [10^{-2}]	$b_1^{\pi}[6]$ (1) [10^{-2}]	$b_1^{\pi}[6]$ (3) [10^{-2}]	b_1^{6q} [10^{-2}]
0.012	0.51	11.20	5.51	2.77	10.5	15.5	24.1	0.00
0.032	1.06	5.50	2.53	1.84	5.6	6.8	8.9	0.00
0.063	1.65	3.82	1.11	0.60	4.2	3.7	4.1	0.00
0.128	2.33	0.29	0.53	0.44	1.6	1.3	1.3	0.01
0.248	3.11	0.29	0.28	0.24	-0.55	.13	0.12	0.41
0.452	4.69	-0.38	0.16	0.03	-0.02	-0.02	-0.022	-0.38

4. Hidden color, 6 quark states

The HERMES experimental result [2] presents an interesting puzzle because it observed a significant negative value of b_1 for $x = 0.45$. At such a value of x, any sea quark effect such as arising from double-scattering or virtual pions is completely negligible. Furthermore, the nucleonic contributions are computed to be very small [3, 4], so one must consider other possibilities. We therefore take up the possibility that the deuteron has a six-quark component that is orthogonal to two nucleons. Such configurations are known to be dominated by the effects of so-called hidden-color states in which two color-octet baryons combine to form a color singlet [8]. Such configurations can be generated, for example, if two nucleons exchange a single gluon leading to a quantum fluctuation involving an color octet and color anti-octet baryon.

In particular, a component of the deuteron in which all 6 quarks are in the same spatial wave function ($|6q\rangle$) can be expressed in terms on nucleon-nucleon NN, delta-delta $\Delta\Delta$ and hidden color components CC as [8]:

$$|6q\rangle = \sqrt{1/9}|N^2\rangle + \sqrt{4/45}|\Delta^2\rangle + \sqrt{4/5}|CC\rangle. \tag{10}$$

This particular state has an 80% probability of hidden color and only an 11% probability to be a nucleon-nucleon configuration. The 80 % cited here is a purely algebraic number that applies only for completely overlapping nucleons. The real question is the probability that the deuteron consists of 6 quarks are in the same spatial wave function, which is denoted here as P_{6q}. A recent review of hidden color phenomena is presented in [9]. In the following, the term $|6q\rangle$ is referred to interchangeably as either a six-quark or hidden color state.

The discovery of the EMC effect [10] caused researchers to consider the effects of such six-quark states [11] in a variety of nuclear phenomena [12, 13, 14]. Furthermore, the possible discovery of such a state as a di-baryon resonance has drawn recent interest [15]. Therefore we propose a model of a hidden-color six-quark components of the s and d-states of the deuteron. We also note that including a six-quark hidden color component of the deuteron does not lead to a conflict with the measured asymptotic d to s ratio of the deuteron [16]. The EMC effect remains the only nuclear effect that has not been explained using conventional (non-quark) dynamics [18, 17, 19].

Our calculation is exploratory, so we use the simplest model possible. We postulate that the S state of the deuteron has a component with 6 quarks in an s state with total angular momentum 1 and isospin 0. Then the D state has a 6-quark component with any one quark in a $d_{3/2}$ state. We define these states by combining 5 s-state quarks into a spin 1/2 component, which couples with either the $s_{1/2}$ or $d_{3/2}$ single-quark state to make a total angular momentum of 1. We therefore write the wave functions of these states for a deuteron of $J_z = H$ as

$$\psi_{j,l,H}(\mathbf{p}) = \sqrt{N_l} f_l(p) \sum_{m_s,m_j} \mathcal{Y}_{jlm_j} \langle jm_j, \frac{1}{2}m_s|1H\rangle, \tag{11}$$

where $l, j = s_{1/2}$ or $d_{3/2}$, N_l is a normalization constant chosen so that $\int d^3p\bar{\psi}_{j,l,H}(\mathbf{p})\gamma^+\psi_{j,l,H}(\mathbf{p}) = 1$, and $\mathcal{Y}_{jlm_j}$ is a spinor spherical harmonic. The matrix element for transition between the $l = 0$ and $l = 2$ states is given by the light-cone distribution

$$F_H(x_{6q}) = \frac{1}{2}\int d^3p\bar{\psi}_{1/2,0,H}(\mathbf{p})\gamma^+\psi_{3/2,2,H}(\mathbf{p})\delta\left(\frac{p\cos\theta + E(p)}{M_{6q}} - x_{6q}\right), \tag{12}$$

where $E(p) = \sqrt{p^2 + m^2}$ with m as the quark mass, and M_{6q} is the mass of the six-quark bag, x_{6q} is the momentum fraction of the six-quark bag carried by a single quark and $x_{6q}M_{6q} = xM$ [11]. Note that $p\cos\theta$ is the third (z) component of the momentum, so that the plus component of the quark momentum is $E(p) + p\cos\theta$. We take $M_{6q} = 2M$ (its lowest possible value) to make a conservative estimate.

The term of interest $b_1(x)$ is given by

$$b_1^{6q}(x) = \frac{1}{2}(2)\left(F_0(x) - F_1(x)\right)P_{6q}, \tag{13}$$

where P_{6q} is the product of the probability amplitudes for the 6-quark states to exist in the deuteron, and the factor of 2 enters because either state can be in the d-wave. Evaluation of F_H using Eq. (11) leads to the result:

$$b_1^{6q}(x) = -\sqrt{\frac{N_0N_2}{2}}\frac{3}{4\pi}\int d^3p f_0 f_2(3\cos^2\theta - 1)\delta\left(\frac{p\cos\theta + E(p)}{M} - x\right)P_{6q}. \tag{14}$$

To proceed further we, specify the wave functions to be harmonic oscillator wave functions. We take $f_2(p) = -p^2R^2e^{-p^2R^2/2}$, $f_0(p) = e^{-p^2R^2/2}$, where R is the radius parameter. This model is specified by only three parameters: R, the quark mass m, and P_{6q}. The key question is whether such a model can reproduce the HERMES data point at $x = 0.452$ without using a value of P_{6q} large enough to conflict with conventional nuclear physics calculations that do not require a non-zero value. In other words, we ask if the hidden color states provide a substantial mechanism to make b_1 non-zero at large values of x.

We adjust the value of P_{6q} to reproduce the data at that point and see how large a value is needed. Here we use a quark mass of 338 MeV [20]. We expect that the 6-quark state should be somewhat larger than that of a nucleon, and therefore choose R to be 1.2 fm. The results of the calculations are not very sensitive to the exact value of R [1].

The main result is that we can reproduce the value of the HERMES high-x point with value of $P_{6q} = 0.15\%$. This is shown in Table I and Figs. 4 and 5 of Ref. [1]. Combining the effects of pions and hidden color states leads to a reasonable reproduction of the HERMES data for its entire range of x. Furthermore, no mechanism other than hidden color is known to contribute to b_1 at large values of x. Given this, reasonable predictions for the JLab experiment are presented in Ref. [1].

5. Sum rule of Close & Kumano [21]

This sum rule states that $\int dx\, b_1(x) = 0$, and is derived by assuming that b_1 is carried entirely by valence quarks. This is analogous to the Gottfried sum rule for the integral of $F_{2p} - F_{2n}$, which assumed that $\bar{u}(x) = \bar{d}(x)$. Various effects of the sea violate the sum rule, and the violations may be more interesting than the sum rule.

Ref. [1] shows that nucleonic, pion and shadowing effects all violate the sum rule, and for nucleonic and pionic contributions by an infinite amount. However, integration of Eq. (14) does yield a zero due to the tensor nature of the integrand. This means that b_1^{6q} must have regions of x for which it is positive and regions of x for which it is negative.

6. Summary

We find that pionic effects are sizable for values of $x < 0.2$ and reproduces the HERMES data in that region. Furthermore, we find that 6-quark hidden color effects can enter at larger values of x, and that the combination of pionic and hidden color effects reproduces HERMES data. Predictions are made for future JLab data. The Close & Kumano sum rule does not hold for all mechanisms other than 6-quark effects. If the sum rule holds, b_1 must take on both positive and negative values. Observing such would provide evidence for 6-quark hidden-color components of the deuteron.

Acknowledgments

This work has been partially supported by U.S. D. O. E. Grant No. DE-FG02-97ER-41014. I thank M. Alberg for useful discussions and M. Aicher [7] for providing a computer program for q^π.

References

[1] Miller G A 2014 *Phys. Rev.* C **89**, 045203
[2] Airapetian A *et al* [HERMES Collaboration] 2005 *Phys. Rev. Lett.* **95**, 242001
[3] Hoodbhoy P, Jaffe R L and Manohar A 1989 *Nuclear Physics* B **312** 571
[4] Khan H and Hoodbhoy P 1991 *Phys. Rev.* C **44** 1219
[5] Slifer K *et al* 2012 Jefferson Laboratory experiment E12-13-011,
[6] Sutton P J, Martin A D, Roberts R G, and Stirling W J 1992 *Phys. Rev.* D **45** 2349
[7] Aicher M, Schafer A., and Vogelsang W 2010 *Phys. Rev. Lett.* **105** 252003
[8] Harvey M (1981) *Nucl. Phys.* A **352** 301
[9] Bakker B L G and Ji C R 2014 *Prog. Part. Nucl. Phys.* **74** 1
[10] Aubert J J *et al* [European Muon Collaboration] 1983 *Phys. Lett.* B **123**, 275
[11] Carlson C E and Havens T J *Phys. Rev. Lett.* **51**, 261
[12] Miller G A 1984 *Phys. Rev. Lett.* **53**, 2008
[13] Koch V, and Miller 1985 *Phys. Rev. C* **31** 602
[14] Miller G A and Gal A 1987 *Phys. Rev. C* **36**, 2450
[15] Bashkanov M, Brodsky S J and Clement H 2013 *Phys. Lett.* B **727**, 438
[16] Guichon P A M and Miller G A 1984 *Phys. Lett.* B **134**, 15
[17] Geesaman D F, Saito K and Thomas A W 1995 *Ann. Rev. Nucl. Part. Sci.* **45**, 337
[18] Arneodo M 1994 *Phys. Rept.* **240**, 301
[19] Hen O, Higinbotham D W, Miller G A, Piasetzky E and Weinstein L B 2013 *Int. J. Mod. Phys.* E **22**, 1330017
[20] Close F E 1979 *An Introduction to Quarks and Partons* (London: Academic Press)
[21] Close F E and Kumano S 1990 *Phys. Rev.* D **42** 2377

Journal of Physics: Conference Series **543** (2014) 012003 doi:10.1088/1742-6596/543/1/012003

The Deuteron Polarized Tensor Structure Function b_1

K Slifer

University of New Hampshire, Department of Physics, 9 Library Way, Durham, NH 03824, USA

E-mail: karl.slifer@unh.edu

Abstract. We describe Jefferson Lab E12-13-011, an inclusive deep inelastic scattering experiment to measure the leading twist deuteron tensor structure function b_1 in the region $0.16 < x < 0.49$ for $0.8 < Q^2 < 5.0$ GeV^2. The experiment has been conditionally approved, contingent on target performance, with A$^-$ physics rating to run with 30 days of 11 GeV incident beam on a tensor polarized solid target in Jefferson Lab's Hall C.

1. Introduction

The deuteron is the simplest nuclear system and in many ways it is as important to understanding bound states in QCD as the hydrogen atom was to understanding bound systems in QED. Unlike it's atomic analogue, our understanding of the deuteron remains unsatisfying both experimentally and theoretically. A deeper understanding of the deuteron's tensor structure will help to clarify how the gross properties of the nucleus arise from the underlying partons. This provides novel information about nuclear structure, quark angular momentum, and the polarization of the quark sea that is not accessible in spin-1/2 targets.

Four independent helicity amplitudes are sufficient to describe virtual Compton scattering from a spin-1/2 target, after requiring parity and time reversal invariance. This number doubles for a spin-1 target, where the spin can be in three states (+, 0, -). This gives rise to a tensor structure which was first discussed for the deuteron for the real photon case by Pais [1], and later in the virtual photon case, by Frankfurt and Strikman [2]. Hoodbhoy, Jaffe and Manohar [3] introduced the notation which we now follow, whereby the tensor structure is described by the four functions b_1, b_2, b_3 and b_4.

Of particular interest is b_1, which measures the extent to which the nuclear ground state deviates from being a simple collection of nucleons [4]; b_1 would vanish if the deuteron was simply a proton and neutron in a relative S state. At low x, shadowing effects are expected to dominate b_1, while at larger $x-$values where b_1 is anticipated to be small in conventional models, measurements provide a clean probe of exotic QCD effects.

There are several different approaches for predicting the $x-$dependent behavior of b_1 including one pion exchange [5], convolution models with relativistic and binding energy corrections [4], covariant relativistic calculations [6], double-scattering effects [7], and the virtual nucleon approximation [8]. All of these models predict a small or vanishing value at moderate x. However, the first measurement [9, 10] of b_1 at Hermes revealed a cross-over to a large negative value in the region $0.2 < x < 0.5$, albeit with relatively large experimental uncertainty. An atomic beam source was used to generate a deuterium gas target with high tensor polarization. The HERA storage ring provided 27.6 GeV positrons incident on the internal gas target.

As displayed in Fig. 1, (gray points) the tensor asymmetry A_{zz} was found to be non-zero at about the two sigma level, with an apparent zero crossing around $x = 0.3$.

1.1. Hidden Color and 6-Quark Bags

Recently, Miller [11] showed that the Hermes data may indicate exotic effects from hidden color 6-quark configurations. See Fig. 1. Critically, it was demonstrated that no conventional nuclear mechanism can reproduce the large-x Hermes data, but that the 6-quark probability needed to do so ($P_{6q} = 0.0015$) is small enough that it does not violate any known conventional nuclear physics. With this in mind, a precise measurement of b_1 could provide a unique and unambiguous signature of hidden color.

1.2. Virtual Nucleon and Light Cone Approximations

Following the approach of Ref. [2], M. Sargsian [8] has calculated the tensor asymmetry A_{zz} for deep inelastic scattering. Within the PWIA approximation in which deuteron parton distributions are generated due to the partons in the proton and neutron, he predicts no significant tensor asymmetry except at very large x (quasi-elastic scattering) in which case the D-wave becomes important. In the approximation in which only the proton-neutron component of the deuteron is taken into account and nuclear parton distributions are generated through the convolution of partonic distributions of the nucleon and deuteron density matrix (see e.g. Refs. [12, 13]), the deuteron structure function b_1 is related directly to the D-partial wave of the deuteron wave function [8, 12]. As a result, this approximation predicts negligible magnitude for b_1 for $x \leq 0.6$ due to the small Fermi momenta involved in the convolution integral. However, the predicted magnitude of b_1 is large at $x \geq 0.7$ where one expects substantial contribution from the D-waves due to the high momentum component of the deuteron wave function involved in the convolution picture of DIS scattering off the deuteron. In this case, b_1 is very sensitive to the relativistic description of the deuteron and its measurement can be used for checking the different approximations of the high momentum component of deuteron wave function.

In the calculation, either the Virtual Nucleon or Light-Cone approximations are used to calculate the tensor polarization for DIS scattering off the deuteron. In both approximations, only the proton-neutron component of the deuteron is taken into account. In the Virtual Nucleon approximation, the covariant scattering amplitude is reduced by estimating the spectator nucleon propagator at its on-energy shell in the lab frame of the deuteron. Within this approximation the baryonic sum rule is satisfied while the momentum sum rule is not. The latter is due to the fact that part of the light cone momentum of the bound virtual nucleon is lost to the unaccounted non-nucleonic degrees of freedom in the deuteron wave function. In the light cone approximation, the scattering amplitude is estimated as the $E + p_z$ pole of the spectator nucleon on the light cone. In this case the wave function is defined on the light-cone reference frame and it satisfies both baryon number and momentum sum rules. For a detailed comparison of these approximations, see Ref. [13].

1.3. The Close-Kumano Sum Rule

Kumano [14] pointed out that the twist-2 structure functions b_1 and b_2 can be used to probe orbital angular momentum. He then extracted the tensor polarized quark and anti-quark distributions from a fit to the Hermes data [9] and found that a non-negligible tensor polarization of the sea is necessary to reproduce the data.

Following the formalism from the parton model in [3], Close and Kumano [15] related the tensor structure function b_1 to the electric quadrupole form factor of the spin-1 target through a sum rule:

$$\int_0^1 dx\, b_1(x) = -\frac{5}{12M^2} \lim_{t\to 0} t\, F_Q(t) + \frac{1}{9}\left(\delta Q + \delta\bar{Q}\right)_s$$

$$= \frac{1}{9}\left(\delta Q + \delta \bar{Q}\right)_s = 0 \tag{1}$$

where $F_Q(t)$ is the electric quadrupole form factor of a spin-1 hadron at the momentum squared t. The Close Kumano (CK) sum rule is satisfied in the case of an unpolarized sea. The authors note that in nucleon-only models, the integral of b_1 is not sensitive to the tensor-polarization of the sea, and consequently the sum rule is always true, even when the deuteron is in a D-state.

The authors of Ref. [4] calculated the first moment of $b_1(x)$ in a version of the convolution model that incorporates relativistic and binding energy corrections. They found a value of $-6.65 \cdot 10^{-4}$, and emphasize that deviations from this will serve as a good signature of exotic effects in the deuteron wave function. Similarly, Ref. [6] predicts $5 \cdot 10^{-4}$ and $3 \cdot 10^{-5}$ for the relativistic and nonrelativistic calculation of Eq. 1, respectively.

A truncated version of Eq. 1 was evaluated by the Hermes [9, 10] experiment and found to be

$$\int_{0.0002}^{0.85} b_1(x)dx = 0.0105 \pm 0.0034 \pm 0.0035, \tag{2}$$

which possibly indicates a breaking of the Close-Kumano sum rule, and consequently a tensor-polarized quark sea. However, since the comparison is only at the two sigma level, more precise data is needed for a true test.

1.4. Angular Momentum Sum Rule for Spin-1 Hadronic Systems

The b_1 structure function is connected with the spin-1 angular momentum sum rule as discussed in Ref. [16]. By examining the energy momentum tensor for the deuteron, the authors showed that it was possible to define an additional sum rule for b_1 (see Eq. 12 in Ref. [16]) where it was shown that the second moment of this quantity is non vanishing, being related to one of the gravitomagnetic deuteron form factors. A measurement of b_1 would provide a unique test of this idea.

It is also important to notice that b_1 singles out the role of the D-wave component in distinguishing coherent nuclear effects through tensor polarized correlations from the independent nucleon's partonic spin structure. A similar role of the D-wave component was also found in the recently proposed spin sum rule where it plays a non-trivial role producing a most striking effect through the spin flip GPD E. An experimental measurement of b_1 would corroborate this scenario.

2. The Jefferson Lab E12-13-011 Experiment

E12-13-011 will perform an inclusive measurement of the deuteron tensor asymmetry A_{zz} in the region $0.16 < x < 0.49$, for $0.8 < Q^2 < 5.0$ GeV^2 and invariant mass $W \geq 1.85$ GeV. A solid polarized ND_3 target will be used, along with the Hall C spectrometers, and an unpolarized beam. The magnetic field of the target will be held constant along the beamline at all times, while the target state is alternated between a polarized and unpolarized state.

This measurement will provide access to the tensor quark polarization, probe hidden color, and provide data needed to test the Close-Kumano sum rule which has been predicted to be satisfied [15] in the absence of tensor polarization in the quark sea.

The experiment was given A^- physics rating and c_1 conditional approval by PAC 40. Efforts to address the c_1 condition to demonstrate 30% tensor polarization in a solid polarized target are underway at the Universities of Virginia and New Hampshire. A total of 30 days of beam time was approved for production data, with an additional 10.8 days of expected overhead.

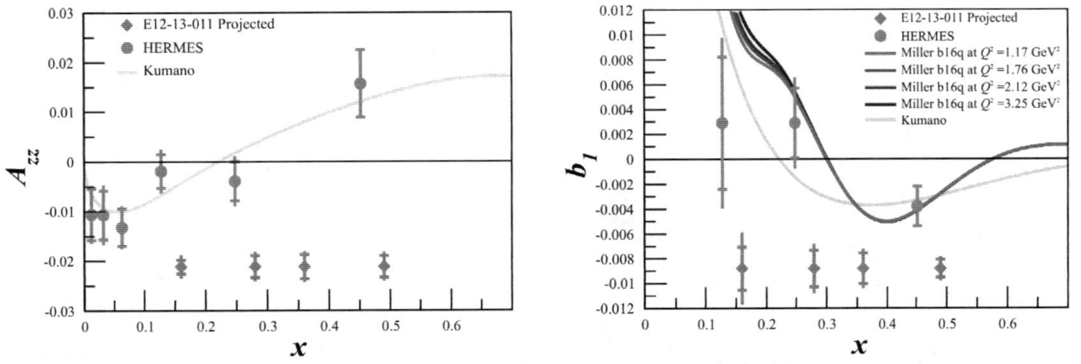

Figure 1. Projected uncertainties for A_{zz} and b_1 assuming a 35% tensor-polarized ND_3 target. The previous data from Hermes [9], a fit from Kumano [14], and the calculation of Miller [11], which assumes a 6-quark probability of $P_{6q} = 0.0015$, are also shown.

2.1. Experimental Method

The measured DIS double differential cross section for a spin-1 target characterized by a vector polarization P_z and tensor polarization P_{zz} is expressed as

$$\frac{d^2\sigma_p}{dxdQ^2} = \frac{d^2\sigma}{dxdQ^2}\left(1 - P_z P_B A_1 + \frac{1}{2} P_{zz} A_{zz}\right), \tag{3}$$

where σ_p (σ) is the polarized (unpolarized) cross section, P_B is the incident electron beam polarization, and A_1 (A_{zz}) is the vector (tensor) asymmetry of the virtual-photon deuteron cross section. This allows us to write the positive polarized tensor, $0 < P_{zz} \leq 1$, asymmetry using unpolarized electron beam as

$$A_{zz} = \frac{2}{P_{zz}}\left(\frac{\sigma_1}{\sigma} - 1\right), \tag{4}$$

where σ_1 is the polarized cross section for

$$P_{zz} = \frac{n_+ - 2n_0 + n_-}{n_+ + n_- + n_0}, \text{ for } n_+ + n_- > 2n_0. \tag{5}$$

Here n_m represents the portion of the ensemble in the m state.

Eq. 4 reveals that the asymmetry A_{zz} compares two different cross sections measured under different polarization conditions of the target, positively tensor polarized and unpolarized. To obtain the relative cross section measurement in the same configuration, the same target cup and material will be used at alternating polarization states (polarized vs. unpolarized), and the magnetic field providing the quantization axis will be oriented along the beamline at all times. This field will always be held at the same value, regardless of the target material polarization state. This ensures that the acceptance remains consistent within the stability (10^{-4}) of the super conducting magnet.

Since many of the factors involved in the cross sections cancel in the ratio, Eq. 4 can be expressed in terms of the charge normalized, efficiency corrected numbers of tensor polarized N_1^c and unpolarized N^c counts,

$$A_{zz} = \frac{2}{fP_{zz}}\left(\frac{N_1^c}{N^c} - 1\right). \tag{6}$$

The dilution factor f corrects for the presence of unpolarized nuclei in the target, and the measured asymmetry is directly related to b_1 through

$$b_1 = -\frac{2}{3} F_1^d A_{zz}. \tag{7}$$

We can determine b_1 with sufficient precision to discriminate between conventional nuclear models and the more exotic behavior hinted at by the Hermes data, as shown in Figure 1. It is important to note that *any* significantly non-zero measured value of b_1 in this region is impossible to explain with conventional nuclear physics. It has also been emphasized that no known mechanism can predict a negative value for b_1 at any x [11]. On the other hand, a zero crossing is absolutely necessary for satisfaction of the Close-Kumano rum rule, so it is compelling to clarify the behaviour of b_1 at large x.

2.2. Uncertainties

We expect scale dependent systematic uncertainties to contribute at the 9% relative level, dominated by the uncertainty arising from target polarimetry (8%) and dilution factor (4%).

Eq. 6 involves the ratio of counts, which leads to cancellation of several first order systematic effects. However, the fact that the two data sets will not be taken simultaneously leads to a sensitivity to time dependent variations which will need to be carefully monitored and suppressed. Full discussion of these effects can be found in Ref. [17].

3. Summary

E12-13-011 will determine b_1 with sufficient precision to discriminate between conventional nuclear models and the more exotic behavior which is hinted at by the Hermes data. The UVa solid polarized ND_3 target will be used, along with the Hall C spectrometers, and an unpolarized beam. An additional 10.8 days will be needed for overhead. This measurement is sensitive to the tensor polarization of the quark sea, allows a test of the Close-Kumano sum rule, and provides a clean probe of hidden color. Until now, tensor structure has been largely unexplored, so the study of these quantities holds the potential of initiating a new field of spin physics at Jefferson Lab.

Acknowledgements

I would like to acknowledge the co-spokespersons of E12-13-011 : Jian-Ping Chen, Narbe Kalantarians, Dustin Keller, Elena Long, Oscar Rondon and Patricia Solvignon.

References

[1] Pais A 1967 *Phys. Rev. Lett.* **19** 544–546
[2] Frankfurt L L and Strikman M I 1983 *Nucl. Phys.* **A405** 557–580
[3] Hoodbhoy P, Jaffe R L and Manohar A 1989 *Nucl. Phys.* **B312** 571
[4] Khan H and Hoodbhoy P 1991 *Phys. Rev.* **C44** 1219–1222
[5] Miller G A 1989 *Proceedings of Electronuclear Physics with Internal Targets* 30–33
[6] Umnikov A Y 1997 *Phys. Lett.* **B391** 177–184
[7] Bora K and Jaffe R L 1998 *Phys. Rev.* **D57** 6906–6911
[8] Sargsian M 2014 *J. Phys.: Conf. Series* Proceedings of the 2014 Tensor Polarized Solid Target Workshop `http://www.jlab.org/conferences/tensor14`
[9] Airapetian A *et al.* (HERMES) 2005 *Phys. Rev. Lett.* **95** 242001
[10] Riedl C 2005 *Ph. D thesis, DESY-THESIS-2005-027*
[11] Miller G A 2014 *Phys.Rev.* **C89** 045203
[12] Frankfurt L and Strikman M 1981 *Phys.Rept.* **76** 215–347
[13] Sargsian M M, Simula S and Strikman M I 2002 *Phys. Rev.* **C66** 024001
[14] Kumano S 2010 *Phys. Rev.* **D82** 017501
[15] Close F and Kumano S 1990 *Phys.Rev.* **D42** 2377–2379

[16] Taneja S K, Kathuria K, Liuti S and Goldstein G R 2012 *Phys.Rev.* **D86** 036008
[17] Wood S 2014 *J. Phys.: Conf. Series* Proceedings of the 2014 Tensor Polarized Solid Target Workshop `http://www.jlab.org/conferences/tensor14`

Journal of Physics: Conference Series **543** (2014) 012004 doi:10.1088/1742-6596/543/1/012004

Hidden Color

C-R Ji

Department of Physics, Box 8202, North Carolina State University, Raleigh, NC 27695-8202

E-mail: ji@ncsu.edu

Abstract. With the acceptance of QCD as the fundamental theory of strong interactions, one of the basic problems in the analysis of nuclear phenomena became how to consistently account for the effects of the underlying quark/gluon structure of nucleons and nuclei. Besides providing more detailed understanding of conventional nuclear physics, QCD may also point to novel phenomena accessible by new or upgraded nuclear experimental facilities. We discuss a few interesting applications of QCD to nuclear physics with an emphasis on the hidden color degrees of freedom.

1. Introduction

The 12 GeV upgrade of the continuous electron beam accelerator facility (CEBAF) in Jefferson Laboratory (JLab) provides an opportunity to investigate novel nuclear phenomena predicted by QCD. In particular, a JLab collaboration proposed to measure the deuteron tensor structure function which may provide a probe of exotic QCD effects due to hidden color in a six-quark configuration[1]. It is also interesting to note the recent measurements of electron scattering from high-momentum nucleons in nuclei performed at Hall C(E02-019) in JLab[2], which may allow for an improved determination of the strength of two- and three-nucleon short-range correlations for several nuclei. This result may indicate the existence of hidden color in multi-quark configurations. These new developments in experimental facilities motivate us to discuss a number of applications of QCD to nuclear physics which is the main concern of nuclear chromodynamics (NCD)[3]. Its goal is to give a fundamental description of nuclear dynamics and nuclear properties in terms of quark and gluon fields at short distance, and to obtain a synthesis at long distances with the normal nucleon, isobar, and meson degrees of freedom. NCD provides an important testing ground for coherent effects in QCD and nuclear effects at the interface between perturbative and non-perturbative dynamics.

NCD may imply in some cases a breakdown of traditional nuclear-physics concepts. For example, one can identify where off-shell effects modify traditional nuclear physics-formulas, such as the impulse approximation for elastic nuclear form factors. At high momentum transfer, nuclear amplitudes are predicted to have a power-law fall off in QCD in contrast to the Gaussian or exponential fall off usually assumed in nuclear physics.

In QCD, the fundamental degrees of freedom of nuclei as well as hadrons are postulated to be the spin-1/2 quark and spin-1 gluon quanta. Nuclear systems are identified as color-singlet composites of quark and gluon fields, beginning with the six-quark Fock component of the deuteron. An immediate consequence is that nuclear states are a mixture of several color representations which cannot be described solely in terms of the conventional nucleon, meson, and isobar degrees of freedom: there must also exist hidden-color multi-quark wave-function

Journal of Physics: Conference Series **543** (2014) 012004 doi:10.1088/1742-6596/543/1/012004

components, i.e., nuclear states which are not separable at large distances into the usual color-singlet nucleon clusters.

Because of color confinement, one may expect that virtually any color-singlet hadronic configuration of quarks and gluons can form either bound states or resonances as discussed in a recent paper by Bashkanov, Brodsky and Clement[4]. Color confinement can lead to $q\bar{q}q\bar{q}$ tetraquark systems[5, 6] such as the charged $Z_c(3900)$[7, 8] and possibly $qqqq\bar{q}$ pentaquark states[1] in addition to the familiar $q\bar{q}$ mesons, qqq baryons, the gg and ggg glueball states[9], as well as nuclei. Recently, the LHCb collaboration has collected a strong enough signal to declare that $Z(4430)$ is a bona fide particle[10], which joins other exotic particles, such as $Z_c(3900)$. The quark content of $Z(4430)$ as well as $Z_c(3900)$ posed a puzzle: its decay implied it contained a charm quark and anticharm, while its charge required two more quarks (a down and anti-up, for example)-giving a total of four[2]. The LHCb experiment at CERN in Geneva, which is primarily set up to study bottom-quark physics in the LHCs proton-proton collisions, has collected 25,000 relevant B_0 decays at energies of 7 and 8 TeVs. This sample is a factor of 10 larger than the data sets of Belle and BaBar. The analysis by the LHCb collaboration shows a highly significant signal (about 14 standard deviations above background) that removes any doubt that $Z(4430)$ is a real particle. The team also confirms that the particle has a spin of 1 and a positive parity, which rules out the interpretation of the particle signatures as merely arising from a pair of (two-quark) D mesons. The only remaining explanation appears that $Z(4430)$ is a bound state of four quarks.

As described in Ref.[4], mesonic nuclei[11, 12, 13, 14, 15, 16, 17] and nuclear-bound quarkonium[18, 19, 20] are also possible. Resonances in the $\bar{q}\bar{q}\bar{q}qqq$ channel just below the $B\bar{B}$ threshold could explain the anomalously large rates[21] seen in $e^+e^- \to p\bar{p}, n\bar{n}, \Lambda\bar{\Lambda}$ at threshold. The anomalously large transverse spin-spin correlation A_{NN} observed in large-angle proton-proton elastic scattering near the strangeness and charm thresholds[22] could be explained by the effects of $|uuduudQ\bar{Q}>$ baryon number $B = 2$ resonances in the $J = L = 1$ pp s-channel[23, 24].

The possible mechanisms underlying confinement multiply as the number of quarks and gluon constituents in a hadronic system increase. A key question is whether such states are bound by fundamental QCD interactions or do the constituents always cluster as color-singlet subsystems? In the case of nuclei, the quark constituents evidently cluster as color-singlet nucleons bound by virtual meson exchange, the analog of covalent binding in molecular physics due to quark interchange or exchange. When there are no covalence quarks in common, QCD also predicts attractive multigluonic van der Waals forces which are due to glueball exchange. The attractive QCD van der Waals potential leads to the prediction of bound states of heavy quarkonium to heavy nuclei[18, 19, 25]. However, there are also rare configurations in which other multiquark color configurations (hidden color [26, 27, 28]) can enter.

As an extreme example, one may expect a huge enhancement of hidden color degrees of freedom in the neutron stars. They are extreme cases of rare isotopes naturally born after going through the stellar explosion process. Neutron stars can be made in the aftermath of type II supernovae explosions which result from the gravitational core collapse of massive stars. While the masses of type II supernovae are greater than $8M_\Theta$, where the solar mass $M_\Theta = 2 \times 10^{33}$ g, the masses of neutron stars are between one and two solar masses[29]. The radius of the netron star is only about 10 km, while the radius of our sun is 6.96×10^5 km. Thus, the central density could reach as high as five to ten times of the normal nuclear density, $\rho_0 = 2.65 \times 10^{14}$ g cm^{-3}. For comparison, the solar density is only $1.4gcm^{-3}$. In such dense nuclei, it appears natural to expect a large enhancement of the hidden color effect. The cooling mechanism of such neutron

[1] The nomenclature $q\bar{q}$ etc. refers to the lowest particle number Fock state of the hadronic eigensolution of the QCD light-front Hamiltonian.

[2] This was presented as one of the Synopses in the website http://physics.aps.org/synopsis-for/10.1103/PhysRevLett.112.222002.

stars has been the subject of numerous studies[30, 31, 32, 33, 35, 36, 37, 38, 39].

On the other hand, it was also claimed in the past that the concept of hidden color is just an artifact of a certain technique of constructing group representations and therefore has no true physical meaning[40]. To understand the main point of such claim, one may consider an example of four-fermion spin system using the SU(2) group instead of jumping into the discussion of the multi-quark color system with the SU(3) group. Consider a system of four spin-1/2 particles and try to find the totally spin-neutral (or spin-singlet) system. If the four-particle system is represented by the two clusters of the two-particle system, then there are two independent representations which provide the totally spin-neutral system: i.e.

$$\begin{aligned} |\mathrm{Singlet\,I}\rangle_{1234} &= \begin{array}{|c|c|}\hline 1 & 3 \\ \hline 2 & 4 \\ \hline \end{array} = |\mathrm{Singlet}\rangle_{12}|\mathrm{Singlet}\rangle_{34}, \\ |\mathrm{Singlet\,II}\rangle_{1234} &= \begin{array}{|c|c|}\hline 1 & 2 \\ \hline 3 & 4 \\ \hline \end{array} = |\mathrm{Triplet}\rangle_{12}|\mathrm{Triplet}\rangle_{34} \\ &= (2|\mathrm{Singlet}\rangle_{13}|\mathrm{Singlet}\rangle_{24} - |\mathrm{Singlet}\rangle_{12}|\mathrm{Singlet}\rangle_{34})/\sqrt{3}. \end{aligned} \tag{1}$$

The key point of the claim in Ref.[40] is that the representation of $|\mathrm{Singlet\,II}\rangle_{1234}$ originally expressed as the product of the two triplet-states can be rewritten as a combination of the entirely singlet products. One should note, however, that the required singlet clusters are not just $|\mathrm{Singlet}\rangle_{12}$ and $|\mathrm{Singlet}\rangle_{34}$ but also $|\mathrm{Singlet}\rangle_{13}$ and $|\mathrm{Singlet}\rangle_{24}$ in order to express $|\mathrm{Singlet\,II}\rangle_{1234}$ in terms of only the singlet-states. In other words, the clustering should be modified in this expression compared to the original clustering of the two triplet states and thus another degree of freedom beyond the $|\mathrm{Singlet}\rangle_{12}|\mathrm{Singlet}\rangle_{34}$ is clearly necessary. The new degree of freedom beyond the $|\mathrm{Singlet}\rangle_{12}|\mathrm{Singlet}\rangle_{34}$ is necessary no matter how one arranges the clustering. In SU(3) color system, such new degrees of freedom beyond the original clustering of singlet-states correspond to the hidden-color states and they are clearly necessary to express all the totally color-singlet states of any multi-quark system beyond the ordinary meson and baryon systems.

In the next section, Section 2, we discuss the hidden-color degrees of freedom along with the first-principle QCD evolution of multi-quark systems. In particular, we pay attention to the six- and nine-quark hidden-color configurations and show how fast the number of hidden-color degrees of freedom increases as the number of quarks increase. In Section 3, we review the work by Bashkanov, Brodsky and Clement[4] for the recent observation of a hadronic resonance d^* in the proton-neutron system with isospin $I = 0$ and spin-parity $J^P = 3^+$ which raises the possibility of producing other novel six-quark dibaryon configurations allowed by QCD. The width and decay properties of such six-quark resonances could be regarded as manifestations of hidden-color six-quark configurations. Finally, in Section 4 we discuss one of the basic problems in the analysis of nuclear scattering amplitudes, namely how to consistently account for the effects of the underlying quark/gluon component structure of nucleons. We review the idea of the reduced nuclear amplitudes and discuss its application to the coherent pion photoproduction on the deuteron. Summary and conclusions follow in Section 6.

2. Multiquark Evolution in QCD and Hidden Color Degrees of Freedom

The short-distance behavior of multi-quark wave functions can be systematically computed in pQCD. A simple illustration may be found in Ref. [27], where the wave function of a four-quark color-singlet bound state in $SU(2)_C$ has been analyzed as an analogue to the six-quark problem in $SU(3)_C$. The QCD evolution equation was solved for the multi-quark distribution amplitude at short distances in the basis of completely anti-symmetrized quark representations. The eigensolutions of the evolution kernel correspond to a spectrum of candidate states of the relativistic multi-quark system. The four-quark antisymmetric representations are then connected to the eigensolutions to the physical two-cluster basis of $SU(2)_C$ dibaryon (NN,

Journal of Physics: Conference Series **543** (2014) 012004 doi:10.1088/1742-6596/543/1/012004

$N\Delta$, $\Delta\Delta$) and hidden-color (CC) components. It provides constraints on the effective nuclear potential between two clusters. Anomalous states are also found in the spectrum which cannot exist without substantial hidden-color degrees of freedom.

As technical details are presented in Ref. [27], a given four-quark antisymmetric representation (A) can be decomposed onto two clusters ($A_1 \otimes A_2$) using the following steps:

(i) Represent the four-quark antisymmetric representation as an inner product form $A = C \times T \times S \times O$.

(ii) Decompose each four-quark representation C, T, S, and O as an outer product of 2 two-quark representations using fractional parentage coefficients, e.g., $C = C_1 \otimes C_2$.

(iii) Recombine the representations as an inner product: $A = (C_1 \otimes C_2) \times (T_1 \otimes T_2) \times (S_1 \otimes S_2) \times (O_1 \otimes O_2)$.

(iv) Commute the order of inner product and outer product, gathering together representations of the same cluster: $A = (C_1 \times T_1 \times S_1 \times O_1) \otimes (C_2 \times T_2 \times S_2 \times O_2) \equiv A_1 \otimes A_2$.

(v) It is sufficient to consider only the coefficient of the symmetric orbitals O_1 and O_2 to classify the clusters such as NN, $N\Delta$ and $\Delta\Delta$.

With this method, the four-quark eigensolutions can be expanded on the physical basis of effective clusters which are the analogs of the NN, $\Delta\Delta$, $N\Delta$, and CC states in QCD. By analyzing the behavior of $\phi(x_i, Q)$ at large Q, one can predict the effective potential between two clusters. For example, it was found that one of the hidden-color states has a large projection on the eigensolution with leading anomalous dimension (dominant at short distances), whereas the states analogous to NN and $\Delta\Delta$ in QCD have an almost negligible leading component. This implies that the effective potential tends to be repulsive between color-singlet clusters and attractive between colored clusters at short distance. Two other types of four-quark states were also found in $SU(2)_C$, which cannot be identified with dibaryon degrees of freedom. One of these states has equal NN, $\Delta\Delta$ and CC components. The other state is an anomalous hidden-color two-cluster system orthogonal to the usual hidden-color state which has the unusual feature that it has very small projection on the eigensolutions which dominate at short distance, i.e., the effective potential between the colorful clusters of the anomalous hidden-color state tends to be repulsive. One may speculate that the analogous anomalous states in QCD could be quasistable non-nucleonic nuclear systems, possibly related to the anomalous phenomena apparently observed in nuclear collisions [41, 42, 43, 44]. These results also give some support to the conjecture that multi-quark hidden-color components exist in ordinary nuclei [43].

The results in Ref. [27] represented a first attempt to extract exact results for the composition and interactions of multi-quark nuclear systems at short distances. Although just the four-quark bound state in $SU(2)_C$ was analyzed in Ref. [27] for simplicity, many of the derived properties were indeed extended to six-quark states in QCD as discussed in Ref. [28]. In particular, since the leading eigensolution at high-momentum transfer has 80% hidden-color probability, we expect a transition of the ordinary nuclear state to non-nucleonic degrees of freedom as one evolves from long to short distances. The set of eigensolutions of the evolution equation represent all the possible degrees of freedom of the multi-quark bound-state system since its kernel has the same invariances and symmetries of the full QCD Hamiltonian. We thus expect that the eigensolutions of the evolution kernel, which are dominantly hidden-color, to correspond to actual states and excitations of ordinary nuclei. A careful experimental search for these exotic resonances should be made. Possible channels where signals for such states may be observed include Compton scattering and pion photo-production on a deuteron target at large angles, as well as the electron scattering from high-momentum nucleons in nuclei.

2.1. Six-Quark Evolution

Six-quark states can be classified by their symmetries under $SU(3)_C$ (color), $SU(2)_T$ (isospin), $SU(2)_S$ (spin), and spatial symmetry. Since the physical states are color singlets, the Young symmetry of the color-singlet states of the six-quark system is [222] or . Since three colors are shared by six quarks, there are five independent color-singlet states corresponding to five different Yamanouchi labels of [222] symmetry. The explicit representations of the five independent color-singlet states and their correspondence to Yamanouchi labels are given in the Appendix of Ref. [28].

Harvey[45, 46, 47] has classified the color-singlet six-quark states in terms of a physical cluster decomposition. In this classification, the physical deuteron state (i.e., a bound state of two color-singlet clusters) is represented as a linear combination of several different kinds of totally antisymmetric color-singlet six-quark states. For example, the two well-separated nucleons $|NN\rangle$ are given by

$$|NN\rangle = \frac{1}{3}|[6]\{33\}\rangle + \frac{2}{3}|[42]\{33\}\rangle - \frac{2}{3}|[42]\{51\}\rangle, \tag{2}$$

where [] and {} represent the orbital and spin-isospin symmetry and color symmetry [222] is abbreviated. Similarly, the two well-separated deltas $|\Delta\Delta\rangle$ and the hidden-color states $|CC\rangle$ are given by

$$|\Delta\Delta\rangle = \sqrt{\frac{4}{45}}|[6]\{33\}\rangle + \sqrt{\frac{16}{45}}|[42]\{33\}\rangle + \sqrt{\frac{25}{45}}|[42]\{51\}\rangle, \tag{3}$$

$$|CC\rangle = \sqrt{\frac{4}{5}}|[6]\{33\}\rangle - \sqrt{\frac{1}{5}}|[42]\{33\}\rangle, \tag{4}$$

respectively.

Since this classification itself does not include the dynamics of strong interactions between the constituents, the dynamical evolution equation of six-quark systems was formulated [28] to include the dynamics between the quarks and solved to give the general form of the quark distribution amplitude $\phi_d(x_i, Q)$:

$$\phi_d(x_i, Q) = (CTS)\phi(x_i, Q), \tag{5}$$

where (CTS) is a tensor representation obtained from the Young symmetry of $SU(3)_C$, $SU(2)_T$, and $SU(2)_S$, and the orbital distribution amplitude is given by

$$\phi(x_i, Q) = x_1x_2x_3x_4x_5x_6 \sum_{n=0}^{\infty} a_n \tilde{\phi}_n(x_i) \left(\ln\left(\frac{Q^2}{\Lambda^2}\right)\right)^{-\gamma_n}. \tag{6}$$

One may then project Eq. (2) to momentum space:

$$\phi_{NN}(x_i, Q) = \frac{1}{3}\phi_{[6]\{33\}}(x_i, Q) + \frac{2}{3}\phi_{[42]\{33\}}(x_i, Q) - \frac{2}{3}\phi_{[42]\{51\}}(x_i, Q). \tag{7}$$

In the limit $Q \to \infty$, the dependence on Q is determined by the leading anomalous dimension; all other terms which have nonleading anomalous dimensions are suppressed by logarithmic damping factors. The orbital symmetry of the eigensolution which has the leading anomalous dimension cannot be [42] but is [6]. This means only the first term of Eq. (7) survives at the large-Q limit. The NN amplitude itself is not sufficient. One can show that an 80% hidden-color state is necessary to saturate the normalization of the six-quark amplitude when six quarks approach the same position in impact space. One may call this new degree of freedom an anomalous state since it does not correspond to the usual nucleonic degrees of freedom of the nucleus. The physical implication of the anomalous state was discussed in the previous toy model analysis [27].

The QCD predictions for high-Q behavior of the deuteron form factor and the form of the deuteron distribution amplitude at short distances were presented in Ref. [26]. The fact that the six-quark state is 80% hidden color at small transverse separation implies that the deuteron form factors cannot be described at large Q by meson-nucleon degrees of freedom alone, and that the nucleon-nucleon potential is repulsive at short distances. Since the basic scale of QCD, $\Lambda_{\rm QCD}$, is phenomenologically of the order of a few hundred MeV or less, QCD predicts a transition from the traditional meson and nucleon degrees of freedom of nuclear physics to quark and gluon degrees of freedom at inter-nucleon separations of a fm or less. In this respect, it may be important to realize the 12 GeV upgrade of JLab as a viable opportunity to investigate novel nuclear phenomena predicted by QCD. As an example, a JLab collaboration proposed to measure the deuteron tensor structure function b_1[1]. This leading twist tensor structure function of spin-1 hadrons provides a tool to study partonic effects, while also being sensitive to coherent nuclear properties in the simplest nuclear system. Although shadowing effects are expected to dominate this structure function at low values of the Bjorken scaling variable x, it may provide a probe of exotic QCD effects due to hidden color in 6-quark configuration at larger values of x. Since the deuteron wave function is relatively well known, any novel effects are expected to be readily observable. An analysis including the hidden-color degrees of freedom appeared recently[48]. Most available models predict a small or vanishing value of b_1 at moderate x. However, the first measurement of b_1 at HERMES revealed a crossover to an anomalously large negative value in the region $0.2 < x < 0.5$, albeit with relatively large experimental uncertainty. The proposal [1] describes an inclusive measurement of the deuteron tensor asymmetry in the region $0.15 < x < 0.45$, for $0.8 < Q^2 < 5.0$ GeV^2. It might be possible to determine b_1 with sufficient precision to discriminate between conventional nuclear models, and the more exotic behaviour which is hinted at by the HERMES data. This measurement will provide access to the tensor quark polarization, and allow a test of the Close-Kumano sum rule[49], which vanishes in the absence of tensor polarization of the quark sea.

2.2. Nine-Quark Color Singlets

In this subsection, we provide the counting of the number of hidden-color states in nine-quark systems. It is well known that the baryon multiplets can be constructed from the three triplets (quarks) in $SU(3)$: i.e.,

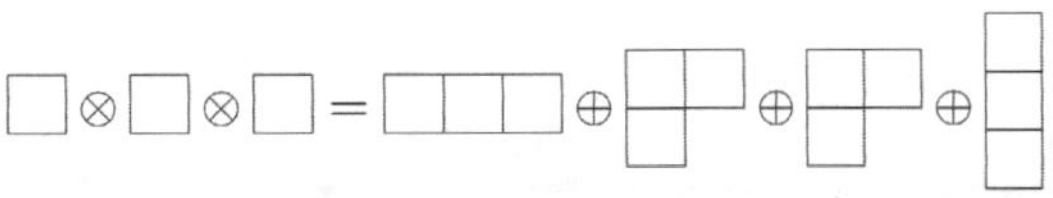
(8)

or

$$\{3\} \otimes \{3\} \otimes \{3\} = \{10\} \oplus \{8\}_{\rm S} \oplus \{8\}_{\rm A} \oplus \{1\}. \tag{9}$$

Applying it in the color degrees of freedom, i.e. $SU(3)_C$, one can see that only one color singlet appears in the three-quark system, or baryon. However, more color singlets independent of each other can be formed as the number of quarks gets increased.

In the six-quark system such as deuteron, five color singlets are formed owing to the fact that the three color degrees of freedom are shared by the six quarks in the system. In other words, not only the singlet and singlet but also the octet and octet can form six-quark color-singlet bound-states. Similar to Eq. (8), one may explicitly get

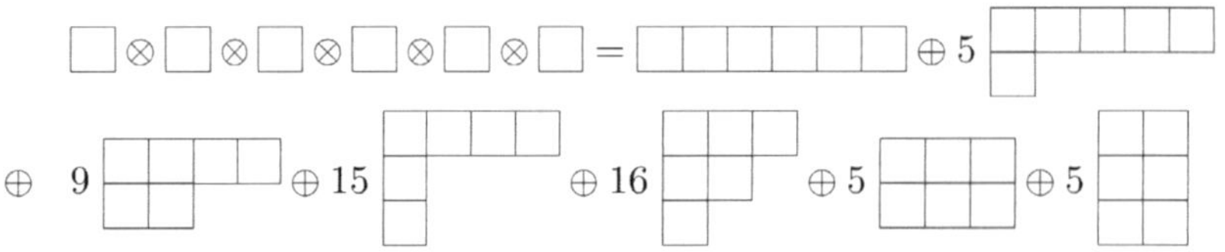

(10)

or

$$
\begin{aligned}
& \{3\} \otimes \{3\} \otimes \{3\} \otimes \{3\} \otimes \{3\} \otimes \{3\} = 729 \\
= \; & \{28\} \oplus 5\{35\} \oplus 9\{27\} \oplus 15\{10\} \oplus 16\{8\} \oplus 5\{10^*\} \oplus 5\{1\}.
\end{aligned} \tag{11}
$$

Among the five color-singlet states in Eq. (11), just one of them stems from the two color-singlet clusters of three-quark system such as N and Δ while the remaining four states correspond to the two color-octet clusters denoted as CC or hidden-color states.

Now, let's count here how many hidden-color states are available in the nine-quark system such as the ^{3}He nucleus. Extending the same color algebra, we get

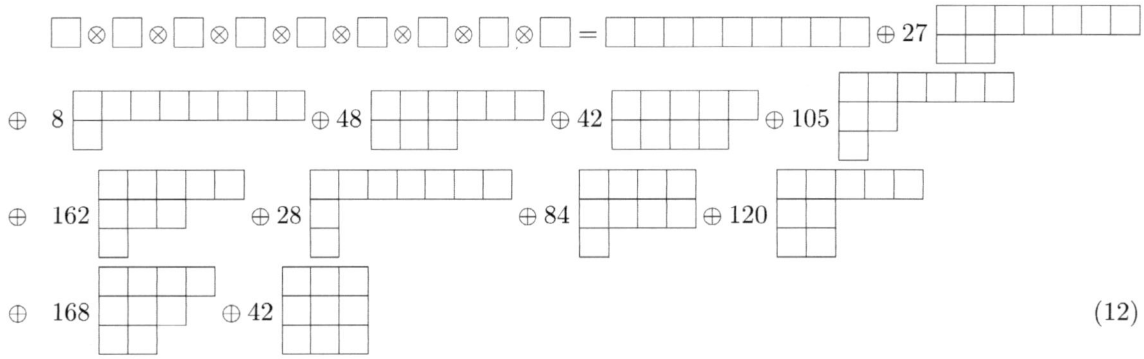

(12)

or

$$
\begin{aligned}
& \{3\} \otimes \{3\} \otimes \{3\} \otimes \{3\} \otimes \{3\} \otimes \{3\} \otimes \{3\} \otimes \{3\} \otimes \{3\} = 19683 \\
= \; & \{55\} \oplus 27\{81\} \oplus 8\{80\} \oplus 48\{64\} \oplus 42\{35^*\} \oplus 105\{35\} \oplus 162\{27\} \oplus 28\{28\} \oplus 84\{10^*\} \\
\oplus \; & 120\{10\} \oplus 168\{8\} \oplus 42\{1\}.
\end{aligned} \tag{13}
$$

This shows a remarkable proliferation of the hidden color degrees of freedom in the nine-quark system compare to the six-quark system[50], i.e. $(42-1)/(5-1) = 41/4 > 10$. Here[3] we find fourty-one hidden-color states in the nine-quark system since only one of the fourty-two color-singlet states stem from the three color-singlet clusters of three-quark systems. Compared to just four hidden-color states in the six-quark system, we now find more than an order of magnitude increase of hidden-color states in the nine-quark system. It amounts to almost an order of magnitude increase in the hidden-color sectors.

It is interesting to note the recent measurements of electron scattering from high-momentum nucleons in nuclei performed at JLab Hall C(E02-019)[2]. These data allowed for an improved determination of the strength of two- and three-nucleon correlations for several nuclei, including

[3] The number of color singlets can also be found from a specific combinatoric counting technique using Young tableaux. The number of hidden-color degrees of freedom presented in Ref.[50] is corrected in this work.

light nuclei where clustering effects can be examined. At $x > 2$, where three-nucleon short-range correlations (3N-SRCs) are expected to dominate, their $A/^3$He ratios were significantly higher than the previous CLAS data and suggested that contributions from 3N-SRCs in heavy nuclei are larger than previously believed. Interestingly, there were large differences between the ^{4}He/^{3}He ratios from E02-019 ($Q^2 \approx 2.9$ GeV^2) and those from CLAS ($\langle Q^2 \rangle \approx 1.6$ GeV^2). If these results indeed indicate a large dependence on Q^2 for the $x > 2$ plateau region, then it might be associated with the dramatic increase of the hidden color degrees of freedom that we found in the three-nucleon systems. It will be interesting to see if this ratio difference between Hall C and CLAS is truly due to the Q^2 difference between these two experiments. This may be tested in the upcoming experiments to be performed after the 12 GeV upgrade in JLab.

3. Double Pionic Fusion and "ABC" Effect

Bashkanov, Brodsky and Clement[4] discussed the recent observation of a hadronic resonance d^* in the proton-neutron system with isospin $I = 0$ and spin-parity $J^P = 3^+$ and the possibility of producing novel six-quark dibaryon configurations allowed by QCD. As Clement reviewed the process of two-pion production in nucleon-nucleon collisions[51], it provides a suitable means to study single as well as mutual nucleon excitations by the well-understood t-channel meson exchange and gives access to study the mutual excitation of the colliding nucleons into their first excited state, the $\Delta(1232)$, in great detail. The question of whether there are more eigenstates in the system of two baryons than just the deuteron ground state has been a strong motivation for this process from the perspective of quantum chromodynamics (QCD) since the possible existence of so-called dibaryons was first envisaged by Jaffe[52]. Although the worldwide dibaryon searches have met with no success so far, the two-pion production is of special interest due to the possible detailed study of baryon-baryon excitations with particular emphasis on the $\Delta\Delta$ system.

In particular, the purely isoscalar reaction channel $pn \to d\pi^0\pi^0$ exhibits a narrow resonance-like structure with $m = 2.37$ GeV and $\Gamma = 70$ MeV in the total cross section, which is correlated with an intriguing low-mass enhancement in the $\pi\pi$ invariant mass spectrum known as the ABC effect[53] as well as with the formation of a $\Delta\Delta$ system in the intermediate state. From the angular distributions, the spin and parity of this structure have been derived resulting in $I(J^P) = 0(3^+)$. For this resonance-like structure to be the ground-state S-wave of the totally antisymmetric six-quark system, it is inevitably represented by[28]

$$|(1^6)\,\yng(1,1,1,1,1,1) >_{CTSO} = \frac{1}{\sqrt{5}} \sum_{Y_T} \sum_{Y_{CS}} \eta_T |\,\yng(3,3)\, Y_T >_T |\,\yng(2,2,2)\, Y_{CS} >_{CS} |\,\yng(6)\,[111111]_O >_O, \qquad (14)$$

where Y_T is the allowed Yamanouchi label of isospin symmetry and the phase $\eta_T = \pm 1 = (-1)^{Y_T}$ depends on whether Y_T is obtained from the Yamanouchi label with the indices in natural order by an even or odd number of transpositions. Similarly, Y_{CS} is the allowed Yamanouchi label of color-spin symmetry and $[111111]_O$ is the only allowed Yamanouchi label for the S-wave orbital symmetry. As phrased long ago in terms of inevitable dibaryon[54], it is interesting to note that an inevitable symmetry is consequential from a particular quantum number such as $I(J^P) = 0(3^+)$.

Such inevitable symmetry in the process of $pn \to d^* \to \Delta\Delta \to d\pi^0\pi^0$ led to the conclusion[4] that d^* must be an unconventional origin, possibly indicating a genuine six-quark nature, since the totally symmetric [6] orbital state corresponds to dominantly the "hidden color" state rather than the "deltaron" state which has the [42] orbital symmetry. From Eqs.(2), (3) and (4), one may note that there are only two possible quark structures for an $I(J^P) = 0(3^+)$ resonance in the two-baryon system given by[4]

$$\begin{aligned} |\psi_{d^*}\rangle &= \sqrt{\frac{1}{5}}|\Delta\Delta\rangle + \sqrt{\frac{4}{5}}|6Q\rangle \quad \text{and} \\ |\psi_{d^*}\rangle &= \sqrt{\frac{4}{5}}|\Delta\Delta\rangle - \sqrt{\frac{1}{5}}|6Q\rangle, \end{aligned} \tag{15}$$

where $\Delta\Delta$ means the asymptotic $\Delta\Delta$ configuration and $6Q$ is the genuine hidden color six-quark configuration. The first solution denotes a S^6 quark structure (all six quarks in the S-shell), or the totally symmetric [6] orbital state, and the second one a S^4P^2 configuration (4 quarks in the S-shell and 2 quarks in the P-shell), or the [42] orbital state. As discussed in Ref.[4], since the quark structure with the large $\Delta\Delta$ coupling would correspond to a "deltaron" state and can be excluded, it may be natural to assign the observed d^* resonance to the S^6 six-quark predominantly hidden color state, thus providing an explanation for its narrow decay width as discussed in Ref.[4].

It is also interesting to note that such a resonance seems to couple with the anomalous enhancement in the low-mass region of the $\pi\pi$-invariant mass spectrum known as the ABC effect[53] and the $\Delta\Delta$ system could well serve as a doorway state to an exotic excitation as discussed by Clement[51]. Such anomalous enhancement has been pointed out already more than fifty years ago in $p + d$ collisions[53]. In the reaction $pd \to {}^3HeX$, an anomalous enhancement in the low-mass region of the $\pi\pi$-invariant mass spectrum was observed[55]. In particular, the missing mass of the detected ${}^3He$ ejectile that corresponds kinematically to the production of a pion pair, i.e. $X = \pi\pi$, exhibited an excess above phase space starting right at threshold at low $\pi\pi$-masses. In subsequent bubble-chamber[56, 57] and single-arm magnetic spectrometer measurements[58, 59, 60, 61, 62, 63, 64], this enhancement was found in fusion reactions leading to d, ${}^3He$ and ${}^4He$, when an isoscalar pion pair was produced. However, no enhancement was found in the fusion reaction to tritium, where an isovector pion pair was produced. These results suggested that the effect only appears in double-pionic fusion reactions in combination with the production of an isoscalar pion pair. In the absence of any consistent explanation, this enhancement effect in the low-mass spectrum was coined as the ABC effect following the initials of the authors of Ref.[53]. For a detailed understanding of the two-pion production process in pp, pd and dd collisions systematic experimental studies were initiated in the nineties at CELSIUS and continued later on at COSY[51]. According to Ref.[51], these exclusive and kinematically complete high-statistics measurements have been carried out at the detector installations PROMICE/WASA[65, 66, 67], CELSIUS/WASA[68, 69, 70, 71, 72, 73, 74], COSY-TOF[75, 76], WASA-at-COSY[77, 78] and COSY-ANKE[79] covering the energy range from the two-pion production threshold up to $\sqrt{s} = 2.5$ GeV (corresponding to a nucleon beam energy of 1.4 GeV).

Since most of these experiments were done at beam energies, which energetically allow the mutual excitation of two colliding nucleons into their first excited state, the $\Delta(1232)P33$, theoretical attempts were undertaken to understand the ABC effect by the excitation of a $\Delta\Delta$ system via t-channel meson exchange between the two colliding nucleons or by variations of this scenario[80, 81, 82, 83, 84]. Such calculations, indeed, predicted a low-mass enhancement, however, always also a high-mass enhancement. Surprisingly enough, the inclusive single-arm measurements at that time seemed to support such a high-mass enhancement. And it was only recently that the first exclusive and kinematically complete measurements of solid statistics[68, 77, 85, 86] over practically the full phase space revealed that the high-mass enhancement observed in the single-arm measurements rather had resulted from other reactions like 3π and η production[63]. However, the new generation of measurements confirmed the

Journal of Physics: Conference Series **543** (2014) 012004 doi:10.1088/1742-6596/543/1/012004

existence of a very pronounced low-mass enhancement. More detailed discussion of the recent theoretical calculations and the experimental strategies for definitive experimental confirmation or exclusion can be found from Ref.[4]. Investigations of the basic double-pionic fusion in the region of the ABC effect has also been discussed in contrast to the conventional fusion processes in the primordial nucleosynthesis and those taking place in star burning[87].

4. Reduced Nuclear Amplitudes

One of the basic problems in the analysis of nuclear scattering amplitudes is how to consistently account for the effects of the underlying quark/gluon component structure of nucleons. Traditional methods based on the use of an effective nucleon/meson local Lagrangian field theory are not really applicable, giving the wrong dynamical dependence in virtually every kinematic variable for composite hadrons. The inclusion of ad hoc vertex form factors is unsatisfactory since one must understand the off-shell dependence in each leg while retaining gauge invariance; such methods have little predictive power. On the other hand, the explicit evaluation of the multi-quark hard-scattering amplitudes needed to predict the normalization and angular dependence for a nuclear process, even at the leading order of QCD coupling constant α_s requires the consideration of millions of Feynman diagrams. Beyond leading order one must include contributions of non-valence Fock states wave functions, and a rapidly expanding number of radiative corrections and loop diagrams. The reduced amplitude method[88], although not an exact replacement for a full QCD calculation, provides a simple method for identifying the dynamical effects of nuclear substructure, consistent with covariance, QCD scaling laws and gauge invariance. The basic idea has already been introduced for the reduced deuteron form factor. More generally, if the nuclear binding is neglected, then the light-cone nuclear wavefunction can be written as a cluster decomposition of collinear nucleons: $\psi_{q/A} = \psi_{N/A} \Pi_N \Psi_{q/N}$ where each nucleon has $1/A$ of the nuclear momentum. A large momentum transfer nucleon amplitude then contains as a factor the probability amplitude for each nucleon to remain intact after absorbing 1/A of the respective nuclear momentum transfer. Each probability amplitude can be identified with the respective nucleon form factor $F(\hat{t}_i = \frac{1}{A^2} t_A)$, where t_A is the square of the transferred momentum to the nucleus with mass number A. Thus for any exclusive nuclear scattering process, the reduced nuclear amplitude can be defined as

$$m = \frac{\mathcal{M}}{\Pi_{i=1}^{A} F_N(\hat{t}_i)}. \tag{16}$$

The predictions of pQCD for pion photoproduction on the deuteron $\gamma D \to \pi^0 D$ have been given[89] at large momentum transfer using the reduced amplitude formalism. The cluster decomposition of the deuteron wave function at small binding only allows the nuclear coherent process to proceed if each nucleon absorbs an equal fraction of the overall momentum transfer. Furthermore, each nucleon must scatter while remaining close to its mass shell. Thus, the nuclear photoproduction amplitude $\mathcal{M}_{\gamma D \to \pi^0 D}(u,t)$ factorizes as a product of three factors: (1) the nucleon photoproduction amplitude $\mathcal{M}_{\gamma N_1 \to \pi^0 N_1}(u/4, t/4)$ at half of the overall momentum transfer, (2) a nucleon form factor $F_{N_2}(t/4)$ at half the overall momentum transfer, and (3) the reduced deuteron form factor $f_d(t)$, which according to pQCD, has the same monopole fall-off as a meson form factor. A comparison with the recent JLAB data for $\gamma D \to \pi^0 D$ of Meekins *et al* [90] and the available $\gamma p \to \pi^0 p$ data [91, 92, 93, 94, 95] shows good agreement between the pQCD prediction and experiment over a large range of momentum transfers and center-of-mass angles. The reduced amplitude prediction is consistent with the constituent counting rule $p_T^{11} \mathcal{M}_{\gamma D \to \pi^0 D} \to F(\theta_{cm})$ at large momentum transfer. This is found to be consistent with measurements for photon lab energies $E_\gamma > 3$ GeV at $\theta_{\rm cm} = 90^\circ$ and 136°.

The predictions of QCD for nuclear reactions are most easily described in terms of light-front (LF) wave functions defined at equal LF time $\tau = t + z/c$ [96]. The deuteron eigenstate can be projected on the complete set of baryon number $B = 2$, isospin $I = 0$, spin $J = 1, J_z = 0, \pm 1$ color-singlet eigenstates of the free QCD Hamiltonian, beginning with the six-quark Fock states. Each Fock state is weighted by an amplitude which depends on the LF momentum fractions $x_i = k_i^+/p^+$ and on the relative transverse momenta $\vec{k}_{\perp i}$. There are five different linear combinations of six color-triplet quarks which make an overall color-singlet, only one of which corresponds to the conventional proton and neutron three-quark clusters. Thus, the QCD decomposition includes four six-quark unconventional states with "hidden color" [28]. Nevertheless, the cluster decomposition theorem [97] states that in the zero binding limit ($B.E. \to 0$), the LF wave function of the deuteron must reduce to a convolution of on-shell color-singlet nucleon wave functions:

$$\begin{aligned} \lim_{B.E.\to 0} \psi^D_{uududd}(x_i, \vec{k}_{\perp i}, \lambda_i) &= \int_0^1 dz \int d^2\ell_\perp \psi^d(z, \ell_\perp) \\ &\times \psi^p_{uud}(x_i/z, \vec{k}_{\perp i} + (x_i/z)\ell_\perp, \lambda_i) \\ &\times \psi^n_{udd}(x_i/(1-z), \vec{k}_{\perp i} - [x_i/(1-z)]\ell_\perp, \lambda_i), \end{aligned} \tag{17}$$

where $\psi^d(z, \ell_\perp)$ is the reduced "body" LF wave function of the deuteron in terms of its nucleon components. Applying this cluster decomposition to an exclusive process involving the deuteron, one can derive a corresponding reduced nuclear amplitude (RNA) [26, 88, 98]. Moreover, at zero binding, one may take $\psi^d(z, \ell_\perp) \to \delta(z - m_p/(m_p + m_n)) \times \delta^2(\ell_\perp)$. In effect, each nucleon carries half of the deuteron's four-momentum.

Thus, in the weak nuclear binding limit, the deuteron form factor reduces to the overlap of nucleon wave functions at half of the momentum transfer, and $F_D(Q^2) \to f_d(Q^2) F_N^2(\frac{Q^2}{4})$ where the reduced form factor $f_d(Q^2)$ is computed from the overlap of the reduced deuteron wave functions [98]. The reduced deuteron form factor resembles that of a particle spin-one meson form factor since its nucleonic substructure has been factored out. The pQCD predicts the nominal scaling $Q^2 f_d(Q^2) \sim const$ [26]. The measurements of the deuteron form factor show that this scaling is in fact well satisfied at spacelike $Q^2 \geq 1$ GeV^2 [99].

Considering a similar analysis of pion photoproduction on the deuteron $\gamma D \to \pi^0 D$ at weak binding, we obtained the reduced amplitude scaling [89]

$$\mathcal{M}_{\gamma D \to \pi^0 D}(u, t) = C' f_d(t) \mathcal{M}_{\gamma N_1 \to \pi^0 N_1}(u/4, t/4) F_{N_2}(t/4), \tag{18}$$

where the constant C' is expected to be close to unity. A comparison with elastic electron scattering then yields the following proportionality of amplitude ratios:

$$\frac{\mathcal{M}_{\gamma D \to \pi^0 D}}{\mathcal{M}_{eD \to eD}} = C' \frac{\mathcal{M}_{\gamma p \to \pi^0 p}}{\mathcal{M}_{ep \to ep}}. \tag{19}$$

The new factored form, Eq. (18), differs significantly from the older reduced nuclear amplitude factorization [88], for which

$$\mathcal{M}^{\text{older}}_{\gamma D \to \pi^0 D}(u, t) \simeq m_{\gamma d \to \pi^0 d}(u, t) F_N^2(t/4). \tag{20}$$

Here, $m_{\gamma d \to \pi^0 d}$ is the reduced amplitude; it scales the same as $m_{\gamma p \to \pi^0 p}$ at fixed angles since the nucleons of the reduced deuteron d are effectively point-like. The advantages of this reduction are that some nonperturbative physics is included via the nucleon form factors and that systematic extension to many nuclear processes is possible [88]. The new factorization given by Eq. (18) is an improvement because it includes nonperturbative effects in the pion production process itself.

JLAB experimental data [90] on π^0 photoproduction from a deuteron target, up to a photon lab energy $E_{\rm lab} = 4$ GeV, were presented as an example inconsistent with both constituent-counting rules (CCR) [100, 101] and old RNA [88] predictions. One potential explanation for this disagreement has been odderon exchange [102, 103]. Because the odderon has zero isospin and is odd under charge conjugation, such an exchange is allowed in the t channel of π^0 photoproduction. However, it is shown in Ref. [89] that the improved factorization given by Eqs. (18) and (19) is in good agreement with the JLAB data [90] for $\gamma D \to \pi^0 D$ and the available $\gamma p \to \pi^0 p$ data [91, 92, 93, 94, 95] as well as the existing $eD \to eD$ and $ep \to ep$ data. There is thus no need to invoke any additional anomalous contribution to understand the data [90].

5. Summary and Conclusion

In summary, we have indicated a few phenomena that go beyond what can be described in terms of the baryon-meson picture of nuclear physics. We have stressed that if one accounts for the substructure of hadrons in terms of quarks and gluons, new phenomena may occur.

We discussed[50] some ideas concerning a modification of the nucleon-meson picture of the nuclear forces. The occurrence of colored multi-quark substructures in nuclei, was found to have consequences for the understanding of the nucleon-nucleon potential at short distances, which means distances of the same order of the size of the nucleons themselves. Two main points were found: (i) the composite nature of the nucleons, which naturally leads to the idea of strong form factors for the meson-nucleon vertices, can be used to estimate these form factors in the quark-gluon picture, and (ii) the occurrence of "exotic" channels, say the colorless combination of two colored clusters, CC, leads in a natural way to an energy-dependent effective, "optical", potential which incorporates the short-range effective repulsion that is responsible for the saturation of the nuclear force as an effect of the leaking of the nucleon-nucleon channel into the CC channel.

Such multi-quark states must have their own evolution as discussed in Sec. 2. Combined with the idea of reduced nuclear amplitudes, this work leads to predictions of the asymptotic behavior of nuclear observables like electro-magnetic form factors, which can be and have been tested in experiments. The deuteron tensor structure function b_1 could be sensitive to hidden color degrees of freedom at large x. The order of magnitude increase in the nine-quark hidden color degrees of freedom may be behind the significant enhancement of the A/^{3}He ratio as Q^2 gets larger. Recent observation of d^* resonance raises the possibility of producing other novel color-singlet six-quark dibaryon configurations allowed by QCD as discussed in Sec. 3.

The link between the traditional nuclear physics and the quark-gluon picture may be provided by the reduced nuclear amplitudes. An elaboration of this idea was given in Sec. 4.

Acknowledgements

I am indebted to a number of my collaborators in completing this work. In particular, I thank Ben Bakker, Stan Brodsky, John Hiller, Jerry Miller and Ziyue Li for their collaborations on the works presented here. This work was supported in part by the DOE Contract No. DE-FG02-03ER41260.

References

[1] Slifer K et al. 2013 The Deuteron Tensor Structure Function b_1 *A Proposal to Jefferson Lab PAC-40 (Update to PR12-11-110)*
[2] Fomin N et al. 2012 *Phys. Rev. Lett.* **108** 092502
[3] Brodsky S J and Ji C R 1986 *Lect. Notes Phys.* **248** 153
[4] Bashkanov M, Brodsky S J and Clement H 2013 *Phys. Lett.* B **727** 438
[5] Karliner M and Nussinov S 2013 *J. High Energy Phys.* **1307** 153
[6] Frishman Y and Karliner M 2013 *J. High Energy Phys.* **1308** 096
[7] Ablikim et al. M 2013 *Phys. Rev. Lett.* **110** 252001
[8] Liu Z Q et al. 2013 *Phys. Rev. Lett.* **110** 252002

[9] Ochs W 2013 *J. Phys.* G **40** 043001
[10] Aaij R et al. (LHCb Collaboration) 2014 *Phys. Rev. Lett.* **112** 222002
[11] Adlarson P et al. 2013 *Phys. Rev. C* **87** 035204
[12] Yokota A, Hiyama E and Oka M 2013 *PTEP 2013* 113D01
[13] Oset E et al. 2012 *Nucl. Phys.* A **881** 127
[14] Friedman E, Gal A and Mares J 2013 Phys. Lett. B **725** 334
[15] Nagahiro H et al. 2013 *Phys. Rev.* C **87** 045201
[16] Bayar M et al. 2012 *Phys. Rev.* C **86** 044004
[17] Kaskulov M et al. 2007 *Phys. Rev.* C **75** 064616
[18] Brodsky S J, Schmidt I A and de Teramond G F 1990 *Phys. Rev. Lett.* **64** 1011
[19] Luke M E, Manohar A V and Savage M J 1992 *Phys. Lett.* B **288** 355
[20] Belyaev V B, Shevchenko N V, Fix A I and Sandhas W 2006 *Nucl. Phys.* A **780** 100
[21] Baldini R, Pacetti S, Zallo A and Zichichi A 2009 *Eur. Phys. J.* A **39** 315
[22] Court G R, Crabb D G, Gialas I, Khiari F. Z., Krisch A D, Lin A M T, Raymond R S, Raylman R R et al. 1986 *Phys. Rev. Lett.* **57** 507.
[23] Brodsky S J and de Teramond G F 1988 *Phys. Rev. Lett.* **60** 1924
[24] Brodsky S J, de Teramond G F and Karliner M 2012 *Annu. Rev. Nucl. Part. Sci.* **62** 1
[25] Yi K 2013 *Int. J. Mod. Phys.* A **28** 1330020
[26] Brodsky S J, Ji C R and Lepage G P 1983 *Phys. Rev. Lett.***51** 83
[27] Brodsky S J and Ji C R 1986 *Phys. Rev.* D **33** 1406
[28] Ji C R and Brodsky S J 1986 *Phys. Rev.* D **34** 1460
[29] Prakash M 1994 The Equation of State and Neutron Stars *Lecture Note for the Winter School on The Equation of State of Nuclear Matter* (Puri, India)
[30] Bahcall J N and Wolf R A 1965 *Phys.Rev.* **140** B1452
[31] Lattimer J M, Pethick C J, Prakash M and Haensel P 1991 *Phys.Rev.Lett.* **66** 2701
[32] Fujii H,Muto T,Tatsumi T and Tamagaki R 1994 *Phys.Rev.* C **50** 3140
[33] Thorsson V, Prakash M, Tatsumi T and Pethick C J 1995 *Phys.Rev.* D **52** 3739
[34] Glashow S L, Iliopoulos J and Maiani L 1970 *Phys.Rev.* D **2** 1285
[35] Prakash M, Prakash M, Lattimer J and Pethick C J 1992 *The Astrophysical Journal* **390** L77
[36] Haensel P and Gnedin O 1994 *Astron.Astrophys.* **290** 458
[37] Voskresenskii D N and Senatorov A V 1986 *Sov.Phys.JETP* **63** 885
[38] Friman B L and Maxwell O V 1979 *The Astrophysical Journal* **232** 541
[39] Ji C R and Min D P 1998 *Phys. Rev.* D **57** 5963
[40] Gonzáles P and Vento V 1986 Hidden Color: A Spurious Concept *Preprint* MIT-CTP-1347
[41] Judek B 1968 *Can. J. Phys.* **46** 343; Judek B 1972 *Can. J. Phys.* **50** 2082
[42] Friedlander E M et al. 1980 *Phys. Rev. Lett.* **45** 1084
[43] Close F 1982 *Nature* **296** 305 (London)
[44] Gabathuler E 1985 *Prog. Part. Nucl. Phys.* **13** 329
[45] Harvey M 1981 *Nucl. Phys.* A **352** 301 [Erratum Harvey M 1988 *Nucl. Phys.* A **481** 834]
[46] Harvey M 1981 *Nucl. Phys.* A **352** 326
[47] Harvey M 1984 *Nucl. Phys.* A **424** 428
[48] Miller G A 2014 *Phys. Rev.* C **89** 045203
[49] Close F E and Kumano S 1990 *Phys. Rev.* D **42** 2377
[50] Bakker B L G and Ji C R 2014 *Prog. Part. Nucl. Phys.* **74**, 1
[51] Clement H 2012 *Prog. Part. Nucl. Phys.* **67** 486
[52] Jaffe R L 1977 *Phys. Rev. Lett.* **38** 195
[53] Abashian A, Booth N E and Crowe K M 1960 *Phys. Rev. Lett.* **5** 258
[54] Goldman T et al. 1989 *Phys. Rev.* C **39** 1889
[55] Booth N E, Abashian A and Crowe K M 1961 *Phys. Rev. Lett.* **7** 35
[56] Bar-Nir I et al. 1973 *Nucl. Phys.* B **54** 17
[57] Abdivaliev A et al. 1979 *Sov. J. Nucl. Phys.* **29** 796
[58] Homer R J et al. 1964 *Phys. Rev. Lett.* **9** 72
[59] Hall J H, Murray T A and Riddiford L 1969 *Nucl. Phys.* B **12** 252
[60] Banaigs J et al. 1973 *Nucl. Phys.* B **67** 1
[61] Banaigs J et al. 1976 *Nucl. Phys.* B **105** 52
[62] Plouin F et al. 1978 *Nucl. Phys.* A **302** 413
[63] Plouin F, Fleury P and Wilkin C 1990 *Phys. Rev. Lett.* **65** 690
[64] Wurzinger R et al. 1999 *Phys. Lett.* B **445** 423
[65] Johanson J et al. 2002 *Nucl. Phys.* A **712** 75
[66] Brodowski W et al. 2002 *Phys. Rev. Lett.* **88** 192301
[67] Patzold J et al. 2003 *Phys. Rev.* C **67** 052202
[68] Bashkanov M et al. 2009 *Phys. Rev. Lett.* **102** 052301
[69] Pauly C et al. 2005 *Nucl. Instrum. Methods* A **547** 294
[70] Skorodko T et al. 2008 *Eur. Phys. J.* A **35** 317
[71] Skorodko T et al. 2009 *Phys. Lett.* B **679** 30
[72] Kren F et al. 2011 *Phys. Lett.* B **702** 312
[73] Skorodko T et al. 2011 *Phys. Lett.* B **695** 115
[74] Skorodko T et al. 2011 *Eur. Phs. J.* A **47** 108
[75] Abd El-Bary S et al. 2008 *Eur. Phys. J.* A **37** 267
[76] Abd El-Samad S et al. 2009 *Eur. Phys. J.* A **42** 159
[77] Adlarson P et al. 2011 *Phys. Rev. Lett.* **106** 242302
[78] Adlarson P et al. 2012 *Phys. Lett.* B **706** 256

[79] Dymov S et al. 2009 *Phys. Rev. Lett.* **102** 192301
[80] Risser T and Shuster M D 1973 *Phys. Lett.* B **43** 68
[81] Bar-Nir I, Risser T and Shuster M D 1975 *Nucl. Phys.* B **87** 109
[82] Anjos J C, Levy D and Santoro A 1973 *Nucl. Phys.* B **67** 37
[83] Gardestig A, Faldt G and Wilkin C 1999 *Phys. Rev.* C **59** 2608
[84] Alvarez-Ruso L 1999 *Phys. Lett.* B **452** 207
[85] Bashkanov M et al. 2006 *Phys. Lett.* B **637** 223
[86] Keleta S et al. 2009 *Nucl. Phys.* A **825** 71
[87] Bashkanov M and Clement H 2013 *Acta Physica Polonica* B **6** 1157
[88] Brodsky S J and Hiller J R 1983 *Phys. Rev.* C **28** 475; Brodsky S J and Hiller J R 1984 *Phys. Rev.* C **30** 412E
[89] Brodsky S J, Hiller J R, Ji C R and Miller G A 2001 *Phys. Rev.* C **64** 055204
[90] Meekins D G et al. 1999 *Phys. Rev.* C **60** 052201
[91] Imanishi A et al. 1985 *Phys. Rev. Lett.* **54** 2497
[92] Shupe M A et al. 1978 *Phys. Rev. Lett.* **40** 271
[93] Bolon G C et al. 1967 *Phys. Rev. Lett.* **18** 926
[94] Anderson R L et al. 1973 *Phys. Rev. Lett.* **30** 627
[95] Anderson R L et al. 1971 *Phys. Rev. Lett.* **26** 30
[96] Brodsky S J, Pauli H C and Pinsky S S 1998 *Phys. Rept.* **301** 299
[97] Namyslowski J M 1980 Clustering: The Necessary And Sufficient Conditions *Preprint* IFT/80/3-WARSAW
[98] Brodsky S J and Ji C R 1986 *Phys. Rev.* D **33**, 2653
[99] Alexa L C et al. 1999 *Phys. Rev. Lett.* **82** 1374
[100] Brodsky S J and Farrar G R 1973 *Phys. Rev. Lett.* **31** 1153; Brodsky S J and Farrar G R 1975 *Phys. Rev.* D **11** 1309
[101] Matveev V A, Muradyan R M and Tavkhelidze A V 1973 *Lett. Nuovo Cim.* **7** 719
[102] Ginzburg I F, Ivanov D Y and Serbo V G 1993 *Phys. Atom. Nucl.* **56** 1474 [*Yad. Fiz.* **56** 45]
[103] Berger E R and Nachtmann O 1999 *Nucl. Phys. Proc. Suppl.* **79** 352

1st Tensor Polarized Solid Target Workshop IOP Publishing
Journal of Physics: Conference Series **543** (2014) 012005 doi:10.1088/1742-6596/543/1/012005

b_1 and the Angular Momentum Sum Rule in the Deuteron

S Liuti and K Kathuria

Department of Physics, University of Virginia, Charlottesville, VA 22904, USA

E-mail: `sl4y@virginia.edu, kk7t@virginia.edu`

Abstract. We formulate the generalized deep inelastic tensor spin structure of the deuteron which can be obtained from deeply virtual Compton scattering and meson production experiments. We discuss its connection to the total quark angular momentum sum rule for a spin-one hadronic system within a gauge invariant decomposition of hadronic spin.

1. Introduction

In Ref.[1] a sum rule for the deuteron angular momentum was derived. The sum rule involves the vector polarization aspects of the deuteron, namely

$$J_q = \frac{1}{2}\int dx\, x\, H_2^q(x,0,0), \tag{1}$$

where $H_2^q(x,\xi,t)$ is one of the five Generalized Parton Distributions (GPDs), H_i, $i = 1,5$, in terms of which the vector component of the generalized deuteron quark-quark correlator is parametrized (x is the struck quark's momentum fraction, ξ is the skewness parameter, t is the four-momentum transfer squared between the initial and final protons) [2]. The first Mellin moment of H_2^q gives the quarks contribution to the deuteron magnetic form factor $G_2(t) \equiv G_M(t)$. Eq.(1) can be compared to the corresponding relation for the nucleon [3],

$$J_q = \frac{1}{2}\int dx\, x\left[H_q(x,0,0) + E_q(x,0,0)\right], \tag{2}$$

where the first moment of the GPD sum, $H_q(x,\xi,t) + E_q(x,\xi,t)$, is the quark contribution to the nucleon magnetic form factor, $F_1(t) + F_2(t) \equiv G_M(t)$. Similarly to the proton GPD E_q, H_2^q does not have a forward partonic limit.

The deuteron spin structure provides a much richer context to study hadronic spin. Its deep inelastic tensor polarized structure is described at leading twist by the structure function $b_1 \equiv H_5(x,0,0)$ [4]. b_1 is equal to zero for two nucleons moving independently from one another. Therefore it can be seen as a more straightforward measure of nuclear modifications in hard processes. An initial experimental measurement from HERMES [5] in the kinematic region $0.01 < x < 0.45$ shows that both b_1 and the integral $\int b_1(x)dx$ are sensibly different from zero, thus suggesting both the presence of non trivial nuclear effects and of tensor polarized sea quarks in the deuteron [6]. In this contribution, we explore the off-forward deuteron tensor polarized

Journal of Physics: Conference Series **543** (2014) 012005 doi:10.1088/1742-6596/543/1/012005

structure with the aim of making a connection to the puzzles associated with the partonic interpretation of hadronic spin. We give a brief overview of the angular momentum sum rule in Section 2, in Section 3 we describe the angular momentum sum rule for the deuteron, in Section 4 we give a formulation of the generalized tensor structure function H_5 in terms of the nucleon components. In particular, our formulation uncovers new contributions from the nucleon helicity flip GPDs in the deuteron. In Section 5 we draw our conclusions.

2. Generalized Parton Distributions and Orbital Angular Momentum

One of the issues driving the research program at Jefferson Labs 12 GeV upgrade is the investigation of the various components of the proton's angular momentum sum rule. Experiments performed since the '80s, have, in fact, confirmed that only about 30% of the proton spin is accounted by quarks spin, and that the quark contribution is dominated by the valence component (see review in [7]). Current efforts, both in theory and experiment, are therefore directed towards determining the contributions of the orbital angular momentum (OAM) of the quarks, as well as of the spin and OAM of the gluons.

The sum rule relating the energy momentum tensor's (EMT) form factors to the nucleon angular momentum components reads [3]

$$\frac{1}{2} = \sum_q J_q + J_g, \tag{3}$$

where J_q and J_g are the quarks and gluons total angular momenta. One can subsequently consider the decomposition

$$J_q = L_q + S_q, \tag{4}$$

where the longitudinal component, $S_q = \frac{1}{2}\Delta\Sigma_q$, is the quark's helicity, while the gluon angular momentum cannot be separated into its spin and orbital components gauge invariantly. In an alternative, precedent formulation [8], it was pointed out that on the light front one can write the longitudinal component as

$$\frac{1}{2} = \frac{1}{2}\Delta\Sigma_q + \mathcal{L}_q + \Delta G + \mathcal{L}_g, \tag{5}$$

where $\Delta\Sigma_q$ is the same quark helicity term appearing in Eq.(4), and the other terms are identified with the quarks orbital angular momentum and the gluons spin and orbital angular momentum, respectively. J_q in Eqs.(3) and (4) is connected to the GPDs which can be observed in deeply virtual compton scattering (DVCS) experiments through Eq.(2). Eq.(1) was derived similarly by working out the connection between the moments of the GPDs appearing in the parametrization of the hadronic tensor for a spin 1 system [2], and the form factors of the EMT [1], which is further discussed in Section 3.

An important issue is how to observe the quarks and gluons orbital angular momentum components in either decomposition from an independent measurement. In [9] L_q was shown to be associated to a twist three GPD,

$$L_q = -\int_{-1}^{1} dx\, G_2(x, 0, 0). \tag{6}$$

On the other hand, several issues remain to be solved before $\mathcal{L}_{q(g)}$ can be uniquely identified with observables.

In this context, exploring the various angular momentum components in the deuteron provides yet another handle on the hadronic spin puzzle. Complementary information can, in fact, be obtained both by accessing different polarization states that are not allowed in a spin 1/2 target, and because of the isoscalar property leading to specific cancellations of the u and d quarks contributions.

Journal of Physics: Conference Series **543** (2014) 012005 doi:10.1088/1742-6596/543/1/012005

3. Angular momentum in a Spin 1 target

In Ref.[1] we connected the deuteron GPDs to angular momentum by considering the parametrization of the matrix element of the deuteron EMT, $T^{\mu\nu}_{q,g}$, in terms of a new set of gravitational form factors,

$$
\begin{aligned}
\langle p'|T^{\mu\nu}|p\rangle &= -\frac{1}{2}P^\mu P^\nu(\epsilon'^*\epsilon)\mathcal{G}_1(t) - \frac{1}{4}P^\mu P^\nu\frac{(\epsilon P)(\epsilon'^* P)}{M^2}\mathcal{G}_2(t) \\
&- \frac{1}{2}\left[\Delta^\mu\Delta^\nu - g^{\mu\nu}\Delta^2\right](\epsilon'^*\epsilon)\mathcal{G}_3(t) - \frac{1}{4}\left[\Delta^\mu\Delta^\nu - g^{\mu\nu}\Delta^2\right]\frac{(\epsilon P)(\epsilon'^* P)}{M^2}\mathcal{G}_4(t) \\
&+ \frac{1}{4}\left[\left(\epsilon'^{*\mu}(\epsilon P) + \epsilon^\mu(\epsilon'^* P)\right)P^\nu + \mu\leftrightarrow\nu\right]\mathcal{G}_5(t) \\
&+ \frac{1}{4}\left[\left(\epsilon'^{*\mu}(\epsilon P) - \epsilon^\mu(\epsilon'^* P)\right)\Delta^\nu + \mu\leftrightarrow\nu\right. \\
&+ \left. 2g_{\mu\nu}(\epsilon P)(\epsilon'^* P) - (\epsilon'^{*\mu}\epsilon^\nu + \epsilon'^{*\nu}\epsilon^\mu)\Delta^2\right]\mathcal{G}_6(t) \\
&+ \frac{1}{2}\left[\epsilon^{*'\mu}\epsilon^\nu + \epsilon'^{*\nu}\epsilon^\mu\right]\mathcal{G}_7(t) + g^{\mu\nu}(\epsilon'^*\epsilon)M^2\mathcal{G}_8(t)
\end{aligned} \tag{7}
$$

where $P = p + p'$ and $\Delta = p' - p$, and $t = \Delta^2$. $\epsilon^\mu(\Lambda), \epsilon'^\mu(\Lambda')$ are the polarization vectors of the deuteron in the initial and final helicity states, respectively.

The various EMT form factors can be then expressed as combinations of second Mellin moments of the deuteron GPDs, proceeding analogously to the spin 1/2 case. The deuteron GPDs were originally identified in [2]. Following Ref.[2], we write the vector component of the deuteron correlator as

$$
\begin{aligned}
W^{\gamma^+}_{\Lambda\Lambda'} &= \int \frac{dz^-}{2\pi}e^{ixP^+z^-}\langle p',\Lambda' \mid \bar{\psi}\left(-\frac{z^-}{2}\right)\gamma^+\psi\left(\frac{z^-}{2}\right)\mid p,\Lambda\rangle\,|_{z^+=0,\mathbf{z}_T=0} \\
&= -(\epsilon'^*\epsilon)H_1 + \frac{(\epsilon n)(\epsilon'^* P) + (\epsilon'^* n)(\epsilon P)}{(Pn)}H_2 - \frac{(\epsilon P)(\epsilon'^* P)}{2M^2}H_3 \\
&+ \frac{(\epsilon n)(\epsilon'^* P) - (\epsilon'^* n)(\epsilon P)}{(Pn)}H_4 + \left\{4M^2\frac{(\epsilon n)(\epsilon'^* n)}{(Pn)^2} + \frac{1}{3}(\epsilon'^*\epsilon)\right\}H_5,
\end{aligned} \tag{8}
$$

where Λ, Λ' are the deuteron helicities, and n is a light-like vector. The first Mellin moments of the deuteron GPDs satisfy the following relations:

$$\int dx H_1(x,\xi,t) = G_C(t) - \frac{2}{3}\frac{-t}{4M_D^2}G_Q, \tag{9a}$$

$$\int dx H_2(x,\xi,t) = G_M(t), \tag{9b}$$

$$\int dx H_3(x,\xi,t) = \frac{1}{1-\dfrac{t}{4M_D^2}}\left[G_M(t) - G_C(t) + \left(1 + \frac{2}{3}\frac{-t}{4M_D^2}\right)G_Q(t)\right], \tag{9c}$$

$$\int dx H_4(x,\xi,t) = 0, \tag{9d}$$

$$\int dx H_5(x,\xi,t) = 0, \tag{9e}$$

where G_C, G_M, and G_Q are the deuteron charge, magnetic and quadrupole form factors. Notice that Eq.(9e) is valid assuming a zero tensor polarization of the antiquarks.

For the second moments, one finds [1]

$$2\int_{-1}^{1} dxx[H_1(x,\xi,t) - \frac{1}{3}H_5(x,\xi,t)] = \mathcal{G}_1(t) + \xi^2\mathcal{G}_3(t), \tag{10a}$$

$$2\int_{-1}^{1} dxxH_2(x,\xi,t) = \mathcal{G}_5(t), \tag{10b}$$

$$2\int_{-1}^{1} dxxH_3(x,\xi,t) = \mathcal{G}_2(t) + \xi^2\mathcal{G}_4(t), \tag{10c}$$

$$-4\int_{-1}^{1} dxxH_4(x,\xi,t) = \xi\mathcal{G}_6(t), \tag{10d}$$

$$2\int_{-1}^{1} dxxH_5(x,\xi,t) = \frac{t}{4M_D^2}\mathcal{G}_6(t) + \mathcal{G}_7(t). \tag{10e}$$

Several relations and observations follow. Eq.(10b) leads to the angular momentum relation, Eq.(1). The total quarks momentum is described in the forward limit by the form factor $\mathcal{G}_1$, which is analogous to the EMT form factor, A_q, in the spin 1/2 case [3]. However, differently from the spin 1/2 case, the form factor $\mathcal{G}_1(t)$ does not simultaneously enter the angular momentum sum rule. Because of parity and time reversal symmetry properties, $\mathcal{G}_7(t)$ is zero at $t = 0$. Therefore we obtain the result

$$\int_{-1}^{1} dxxH_5(x,0,0) \equiv \int dxxb_1(x) = 0 \tag{11}$$

for the second moment of b_1. Although initial experimental results from HERMES are consistent with the presence of a node in $xb_1(x)$ [5], this sum rule could be verified with future dedicated experiments in the large x region [10].

4. Off-forward tensor polarized deuteron structure

The correlator in Eq.(8) is expressed in terms of kinematical variables defined in the "symmetric frame", where we define: $\overline{P} = (p + p')/2$, the average proton momentum, and $\Delta = p - p'$. P is along the z-axis with momentum, $P_3 \approx P^+$. In the symmetric frame, the deuteron polarization vectors are obtained by rotating $\epsilon_\mu^\pm \equiv 1/\sqrt{2}(0; \pm 1, -i, 0)$ to the planes transverse to p and p', respectively. Connecting the correlator in Eq.(8) to the helicity amplitudes for DVCS from a spin-1 hadron allows one to identify observables that would give access to the various deuteron GPDs, and, in particular, the tensor polarized structure function.

The amplitudes for photon-deuteron scattering used in DVCS are given by,

$$f_{\Lambda_\gamma\Lambda,\Lambda_{\gamma'},\Lambda'} = \sum_{\lambda,\lambda'} g_{\lambda\lambda'}^{\Lambda_\gamma\Lambda'_\gamma}(x,\xi,t,Q^2) \otimes A_{\Lambda'\lambda',\Lambda\lambda}(x,\xi,t),, \tag{12}$$

where Λ_γ denotes the helicity of the initial (virtual) photon, $\Lambda_{\gamma'}$ denotes the helicity of the outgoing photon; $\Lambda(\Lambda')$ the helicity of the incoming (outgoing) deuteron; $\lambda(\lambda')$ the helicity of the initial (final) quark. In Eq.(12) we describe the factorization into the hard scattering part, $g_{\Lambda_\gamma\Lambda'_\gamma}^{\lambda\lambda'}$ for the partonic subprocess $\gamma^* + q \to \gamma + q$, and the quark-proton helicity amplitudes, $A_{\Lambda',\lambda';\Lambda,\lambda}$ that contain the GPDs.

The subprocess amplitude is given by

$$g_{\lambda,\lambda'}^{\Lambda_\gamma,\Lambda'_\gamma}(x,\xi) = [\bar{u}(k',\lambda')\gamma^\mu\gamma^+\gamma^\nu u(k,\lambda)] \left(\frac{\epsilon_\mu^{\Lambda_\gamma} \epsilon_\nu^{*\Lambda'_\gamma}}{\hat{s} - i\epsilon} + \frac{\epsilon_\mu^{*\Lambda'_\gamma} \epsilon_\nu^{\Lambda_\gamma}}{\hat{u} - i\epsilon} \right) q^-, \tag{13}$$

where, in the Bjorken limit, $(k+q) \approx \gamma^+ q^-$, $\hat{s} = (k+q)^2 \approx Q^2(X-\zeta)/\zeta$ and $\hat{u} = (k'-q)^2 \approx Q^2 X/\zeta$, and $q^- \approx (Pq)/P^+ = Q^2/(2\zeta P^+)$, with $\zeta = 2\xi/(1+\xi)$, and $X = (x+\xi)/(1+\xi)$.

For DVCS one can consider either the sum over the transverse helicities of the initial photon (unpolarized), or the difference of the helicities in which case the transverse photon is polarized. The outgoing photon is on-shell, thus purely transverse or helicity ± 1. So the leading incoming virtual photon will have the same helicity in the collinear limit. Only $g^{+,+}_{+,+} (= g^{-,-}_{-,-}$ via Parity conservation) for the direct $\hat{s}$ pole term or $g^{-,-}_{+,+} (= g^{+,+}_{-,-})$ for the crossed $\hat{u}$ pole term will survive. For either allowed combination, $g^S_{++} = g^{++}_{++} + g^{--}_{++}$ and $g^A_{++} = g^{++}_{++} - g^{--}_{++}$ such that

$$g^{S(A)} = g^{++}_{++} \pm g^{--}_{++} \quad = \quad \sqrt{X(X-\zeta)} \left(\frac{1}{X-\zeta+i\epsilon} \pm \frac{1}{X-i\epsilon} \right). \tag{14}$$

From the convolution in Eq.(12) we obtain the following decomposition of the transverse photon helicity amplitudes, that we divide into symmetric (S) and antisymmetric (A) components:

$$\begin{aligned}
f^S_{++} &= g^S_{++} \otimes (A_{++,++} + A_{+-,+-}) && \text{(15a)}\\
f^S_{+-} &= g^S_{++} \otimes (A_{-+,++} + A_{++,-+}) && \text{(15b)}\\
f^S_{00} &= g^S_{++} \otimes 2\, A_{0+,0+} && \text{(15c)}\\
f^S_{+0} &= g^S_{++} \otimes (A_{++,0+} + A_{+-,0-}) && \text{(15d)}\\
f^S_{0+} &= g^S_{++} \otimes (A_{0+,++} + A_{0-,+-}) && \text{(15e)}
\end{aligned}$$

and

$$\begin{aligned}
f^A_{++} &= g^A_{++} \otimes (A_{++,++} - A_{+-,+-}) && \text{(16a)}\\
f^A_{+-} &= g^A_{++} \otimes (A_{-+,++} - A_{++,-+}) && \text{(16b)}\\
f^A_{+0} &= g^A_{++} \otimes (A_{++,0+} - A_{+-,0-}) && \text{(16c)}\\
f^A_{0+} &= g^A_{++} \otimes (A_{0+,++} - A_{0-,+-}). && \text{(16d)}
\end{aligned}$$

The amplitudes $A_{\Lambda'\lambda',\Lambda\lambda}$ implicitly contain an integration over the unobserved quark's transerve momentum, k_T, and are functions of $x_{Bj} = Q^2/2M\nu \approx \zeta$, t and Q^2. The convolution integral in Eq.(12) is given by $\otimes \rightarrow \int_{-1}^{1} dx$.

The connection with the correlator is carried out by considering,

$$A_{\Lambda'\lambda',\Lambda\lambda} = \int \frac{dz^-}{2\pi} e^{ixP^+z^-} \left. \langle p', \Lambda' \mid \mathcal{O}_{\lambda'\lambda}(z) \mid p, \Lambda \rangle \right|_{z^+=0, \mathbf{z}_T=0}, \tag{17}$$

$$\mathcal{O}_{\pm\pm}(z) \quad = \quad -i\bar{\psi}\left(-\frac{z}{2}\right) \gamma^+ (1 \pm \gamma_5) \psi\left(\frac{z}{2}\right) \tag{18}$$

The vector correlator, $W^{\gamma^+}_{\Lambda\Lambda'} = V_{\Lambda\Lambda'}$, is given by the symmetric combinations $A_{\Lambda'+,\Lambda+} + A_{\Lambda'-,\Lambda-}$

Journal of Physics: Conference Series **543** (2014) 012005 doi:10.1088/1742-6596/543/1/012005

corresponding to Eqs.(15), namely,

$$V_{++} \equiv f^S_{++} = H_1 - \frac{\Delta_T^2}{2M^2}\frac{1}{1-\xi^2}H_3 - \frac{1}{3}H_5 \tag{19}$$

$$V_{+-} \equiv f^S_{+-} = -\frac{\Delta_T^2}{2M^2}\frac{1}{1-\xi^2}H_3 \tag{20}$$

$$\begin{aligned} V_{00} \equiv f^S_{00} &= -\frac{1+\xi^2}{1-\xi^2}\left(-1+\frac{\Delta_T^2}{4M^2}\right)H_1 + \left[2(1-\xi^2)+2\left(-1+\frac{\Delta_T^2}{4M^2}\right)\frac{1+\xi^2}{1-\xi^2}\right]H_2 \\ &+ \left[(1-\xi^2)+2\frac{1+\xi^2}{1-\xi^2}\left(-1+\frac{\Delta_T^2}{4M^2}\right)+\left(-1+\frac{\Delta_T^2}{4M^2}\right)^2\frac{1}{1-\xi^2}\right]H_3 \\ &+ \left[2\left(-1+\frac{\Delta_T^2}{4M^2}\right)\frac{4\xi}{1-\xi^2}\right]H_4 + \left[(1-\xi^2)+\frac{1}{3}\frac{1+\xi^2}{1-\xi^2}\left(-1+\frac{\Delta_T^2}{4M^2}\right)\right]H_5 \end{aligned} \tag{21}$$

$$\begin{aligned} V_{+0} \equiv f^S_{+0} &= -\frac{\Delta_T}{2M}\frac{1+\xi}{1-\xi}H_1 + \frac{\Delta_T}{2M}\frac{4\xi}{1-\xi^2}H_2 - \frac{1}{2}\frac{\Delta_T}{1-\xi}\left[M(1+\xi)+M\left(1+\frac{\Delta_T^2}{4M^2}\right)\frac{1}{1+\xi}\right]H_3 \\ &- \frac{\Delta_T}{2M}\frac{1+\xi^2}{1-\xi^2}H_4 - \frac{1}{3}\frac{\Delta_T}{2M}\frac{1+\xi}{1-\xi}H_5 \end{aligned} \tag{22}$$

$$V_{0+} \equiv f^S_{+0} = \frac{\Delta_T}{2M}\frac{1-\xi}{1+\xi}H_1 + \frac{1}{2}\frac{\Delta_T}{1+\xi}\left[M(1-\xi)+M\left(1+\frac{\Delta_T^2}{4M^2}\right)\frac{1}{1-\xi}\right]H_3 + \frac{1}{3}\frac{\Delta_T}{2M}\frac{1-\xi}{1+\xi}H_5. \tag{23}$$

In the forward limit ($\xi = 0$, $t = 0$), these expressions reduce to

$$V_{++} \equiv f^S_{++} = H_1 - \frac{1}{3}H_5 \tag{24}$$

$$V_{+-} \equiv f^S_{+-} = 0 \tag{25}$$

$$V_{00} \equiv f^S_{00} = H_1 + \frac{2}{3}H_5 \tag{26}$$

$$V_{+0} \equiv f^S_{+0} = 0 \tag{27}$$

$$V_{0+} \equiv f^S_{0+} = 0. \tag{28}$$

From the above relations, we can solve for H_5, in terms of the helicity structures,

$$H_5(x,0,0) = V_{00} - V_{++}. \tag{29}$$

To explain the partonic structure of the deuteron helicity amplitudes we start from a picture in terms of bound nucleons. The quark-deuteron helicity amplitudes appearing in Eq.(15) are now labeled as $A^D_{\Lambda'\lambda'_q,\Lambda\lambda_q}$. We define them in terms of a convolution of quark-nucleon amplitudes, $A^N_{\lambda'_N\lambda'_q,\lambda_N\lambda_q}$, and the nucleon-deuteron amplitudes, $B_{\Lambda'\lambda'_N,\Lambda\lambda_N}$, where the integral is taken over the unobserved nucleon momenta as

$$A^D_{\Lambda'\lambda'_q,\Lambda\lambda_q} = \sum_{N=n,p}\sum_{\lambda_N,\lambda'_N} B_{\Lambda'\lambda'_N,\Lambda\lambda_N} \otimes A^N_{\lambda'_N\lambda'_q,\lambda_N\lambda_q}, \tag{30}$$

with $\Lambda, \lambda_N, \lambda_q$, being the deuteron, nucleon, and quark helicities, respectively (see Figure 1).

Journal of Physics: Conference Series **543** (2014) 012005 doi:10.1088/1742-6596/543/1/012005

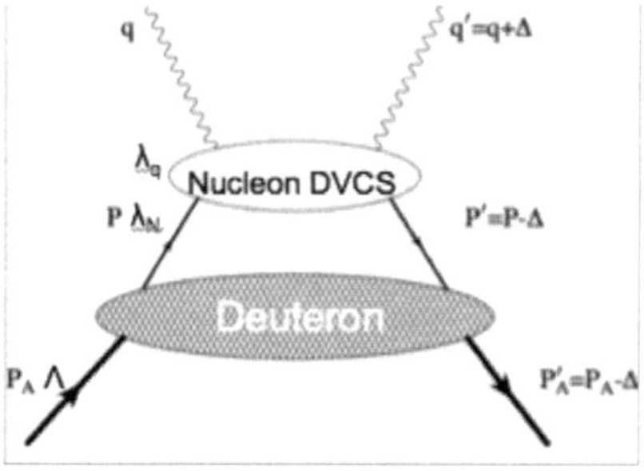

Figure 1. DVCS diagram for scattering off the deuteron in the Impulse Approximation.

We consider the amplitudes governing H_5, namely,

$$\begin{aligned} V_{++} &= A^D_{++,++} + A^D_{+-,+-} = \sum_{N=n,p} \sum_{\lambda_N,\lambda'_N} B_{+\lambda'_N,+\lambda_N} \otimes (A^N_{\lambda'_N+,\lambda_N+} + A^N_{\lambda'_N-,\lambda_N-}) \\ &= \sum_{N=n,p} \left[(B_{++,++} + B_{+-,+-}) \otimes (A^N_{++,++} + A^N_{+-,+-}) \right. \\ &+ \left. (B_{++,+-} + B_{+-,++}) \otimes (A^N_{++,-+} + A^N_{+-,--}) \right] \end{aligned} \quad (31)$$

$$\begin{aligned} V_{00} &= A^D_{0+,0+} = \sum_{N=n,p} \sum_{\lambda_N,\lambda'_N} B_{0\lambda'_N,0\lambda_N} \otimes (A^N_{\lambda'_N+,\lambda_N+} + A^N_{\lambda'_N-,\lambda_N-}) \\ &= \sum_{N=n,p} B_{0+,0+} \otimes (A^N_{++,++} + A^N_{+-,+-}) \end{aligned} \quad (32)$$

The deuteron helicity amplitudes are written as,

$$B_{\Lambda'\lambda'_N,\Lambda\lambda_N} = \sum_{\lambda_{N_2}} \chi^{*\lambda'_N\lambda_{N_2}}_{\boldsymbol{\Lambda}'}(z',p')\chi^{\lambda_N\lambda_{N_2}}_{\boldsymbol{\Lambda}}(z,p), \quad (33)$$

where λ_{N_1} (λ'_{N_1}) are the initial (returning) nucleons' helicities, λ_{N_2} is the spectator nucleon one, and $\chi_\Lambda^{\lambda_{N_1},\lambda_{N_2}}(z,p)$ is the deuteron wave function [11, 12],

$$\chi_\Lambda^{\lambda_{N_1},\lambda_{N_2}}(z,p) = \mathcal{N} \sum_{L,m_L,m_S} \begin{pmatrix} j_1 & j_2 & 1 \\ \lambda_{N_1} & \lambda_{N_2} & m_S \end{pmatrix} \begin{pmatrix} L & S & J \\ m_L & m_S & \Lambda \end{pmatrix} Y_{Lm_L}\left(\frac{\mathbf{p}}{p}\right) u_L(p). \quad (34)$$

In Eq.(34), $j_1 = j_2 = 1/2$, $S = J = 1$. Y_{Lm_L}, where $L = 0, 2$, can be written in terms of the nucleon's light cone variables, consistently with the formalism for describing deep inelastic processes from nuclear targets [13] in the approximation where the quarks' $k_\perp$ dependence is trivially integrated over, and no off-shell effects are considered [14].

By evaluating the various expressions in Eqs.(31,32) one has that:

- $(A^N_{++,++} + A^N_{+-,+-}) \propto H^N$ and $(A^N_{++,-+} + A^N_{+-,--}) \propto E^N$.
- $(B_{++,++} + B_{+-,+-})$ contains both the S and D wave contributions, while $B_{0+,0+}$ contains only S wave contributions. Both terms multiply the same GPD, H^N.
- A new term $(B_{++,+-} + B_{+-,++})$, describing the bound nucleon's helicity flip, and multiplying the GPD E^N, appears in the off-forward configuration.

Therefore in the forward limit, due to the cancellation of the S wave contributions between the V_{++} and V_{0+} terms, $b_1 \equiv H_5(x,0,0)$ is solely proportional to the D wave contribution in the deuteron light cone wave function.

However, in off-forward kinematics we predict a modification and possible enhancement of the tensor structure function, $H_5(x,\xi,t)$ because of the appearance of the nucleon spin flip term $\propto E^N$ in V_{++}.

5. Conclusions

In conclusion, the study of deuteron polarization observables in the exclusive channels can offer unique insights in the context of the proton spin puzzle where observables are currently being sought for possible direct measurements of the quarks and gluons orbital angular momentum components.

In this contribution we discussed the relations between the deuteron energy momentum tensor form factors and GPDs. The angular momentum involves the GPD H_2 that is related to the magnetic properties of the deuteron. Therefore, measurements of the angular momentum sum rule in the deuteron involve DVCS type observables sensitive to the vector polarization.

We also derived a relation for the second moment of the tensor structure function: $xb_1(x)$ integrates to zero at $t = 0$, thus suggesting the presence of a node in this function. Furthermore, we analyzed the helicity structure of $H_5(x,\xi,t)$ and explained that while the dominance of the D wave contribution for this function persists in the off-forward limit, we also uncover a new term that would not be present in inclusive scattering, and is proportional to the bound nucleon spin flip amplitude. Different theoretical scenarios can be envisaged. If this term turns out to be large, it will open investigations of the off-forward tensor structure of the deuteron beyond the D wave component of its wave function. If instead it is small as a consequence of the deuteron isoscalarity ($E_N = E^p + E^n$ is also expected to be small), this would trigger a different viewpoint from where to investigate the off-forward deuteron tensor structure, which would be directly sensitive to the gluon contributions (see also *e.g.* [15]).

A future program of measurements of polarized observables in DVCS from the deuteron stands therefore as an exciting possibility.

Acknowledgments

This work was supported by the U.S. Department of Energy grant DE-FG02-01ER4120.

References

[1] Taneja S K, Kathuria K, Liuti S and Goldstein G R 2012 *Phys. Rev. D* **86** 036008
[2] Berger E R, Cano F, Diehl M and Pire B 2001 *Phys. Rev. Lett.* **87** 142302
[3] Ji X D 1997 *Phys. Rev. Lett.* **78** 610
[4] Hoodbhoy P, Jaffe R L and Manohar A 1989 *Nucl. Phys. B* **312** 571
[5] Airapetian A *et al* 2005 *Phys. Rev. Lett.* **95** 242001
[6] Close F E and Kumano S 1990 *Phys. Rev. D* **42** 2377
[7] Burkardt M, Miller A, and Nowak W D 2010 *Rept. Prog. Phys.* **73** 016201
[8] Jaffe R L and Manohar A 1990 *Nucl. Phys. B* **337** 509
[9] Penttinen M, Polyakov M V, Shuvaev A G, and Strikman M 2000 *Phys. Lett. B* **491** 96; Kiptily D V and Polyakov M V 2004 *Eur. Phys. J. C* **37** 105
[10] Long E *et al* 2014 JLab Letter of Intent LOI12-14-002
[11] Garcon M and Van Orden J W 2001 *Adv. Nucl. Phys.* **26** 293
[12] Gilman R A and Gross F 2002 *J. Phys.* **G28** R37
[13] Ciofi degli Atti C and Liuti S 1990 *Phys. Rev. C* **41**, 1100
[14] Liuti S, Taneja S K 2005 *Phys. Rev. C* **72** 034902
[15] Brodsky S J and Gardner S 2006 *Phys. Lett. B* **643** 22

Journal of Physics: Conference Series **543** (2014) 012006 doi:10.1088/1742-6596/543/1/012006

Final-state interactions in deep-inelastic scattering from a tensor polarized deuteron target

W Cosyn[1] **and M Sargsian**[2]

[1] Department of Physics and Astronomy, Ghent University, Proeftuinstraat 86, 9000 Gent, Belgium

[2] Department of Physics, Florida International University, Miami, Florida 33199, USA

E-mail: wim.cosyn@ugent.be

Abstract. Deep-inelastic scattering (DIS) from a tensor polarized deuteron is sensitive to possible non-nucleonic components of the deuteron wave function. To accurately estimate the size of the nucleonic contribution, final-state interactions (FSIs) need to be accounted for in calculations. We outline a model that, based on the diffractive nature of the effective hadron–nucleon interaction, uses the generalized eikonal approximation to model the FSIs in the resonance region, taking into account the proton-neutron component of the deuteron. The calculation uses a factorized model with a basis of three resonances with mass $W < 2$ GeV as the relevant set of effective hadron states entering the final-state interaction amplitude for inclusive DIS. We present results for the tensor asymmetry observable A_{zz} for kinematics accessible in experiments at Jefferson Lab and Hermes. For inclusive DIS, sizeable effects are found when including FSIs for Bjorken $x > 0.2$, but the overall size of A_{zz} remains small. For tagged spectator DIS, FSIs effects are largest at spectator momenta around 300 MeV and for forward spectator angles.

1. Introduction

Deep-inelastic scattering (DIS) from a polarized spin 1 target yields access to new structure functions not present for the spin 1/2 case [1]. The leading twist structure function b_1 is accessible in DIS from a tensor polarized spin 1 target with an unpolarized beam. In the parton model, b_1 is sensitive only to the spin of the embedding spin 1 hadron. For a spin 1 system of two non-interacting spin 1/2 particles at rest, b_1 is trivially zero. For the deuteron, due to the D-wave admixture in the wave function of a few percent, b_1 is still expected to be small ($b_1 \ll F_1^D$) in the plane-wave approximation. At low Bjorken $x < 0.1$, shadowing effects are expected to yield significant contributions to b_1 [2, 3, 4]. Measurement of a sizeable value for b_1 at moderate x would provide insight into possible non-nucleonic components in the deuteron [5]. The Hermes collaboration measured b_1 for $0.01 < x < 0.45$ at $0.5 < Q^2 < 5$ GeV^2, finding non-zero values and the measured tensor asymmetry A_{zz} [see Eq. (8) below] exhibited a sign change around $x \approx 0.3$ [6]. An experiment at Jefferson Lab (JLab) will provide a new measurement for b_1 at $0.16 < x < 0.5$ and $1 < Q^2 < 5$ GeV^2 [7].

So far, no study has been made of possible final-state interaction (FSI) effects at moderate x to the nucleonic contribution of b_1. It is important to quantify this contribution as a sizeable FSI effect could dominate possible non-nucleonic contributions. Here, we present work in a model that has been used to account for FSIs in DIS from the deuteron for inclusive [8] and tagged spectator measurements [9, 10]. The major difficulty in describing FSI in these reactions is that one lacks detailed information about the composition and space-time evolution of the hadronic system produced in the deep inelastic scattering and how this changes as a function of Bjorken x and Q^2.

2. Formalism

The model introduced in Ref. [9] accounted for FSI effects in DIS from the deuteron based on general properties of high-energy diffractive scattering. The underlying assumption was that due to the restricted phase space (finite values of W and Q^2), the minimal Fock state component of the wave function can be used to describe DIS from the bound nucleon. In this case the scattered state consists of three outgoing valence quarks whose rescattering from the spectator nucleon is parametrized in the form of a Q^2- and W-dependent diffractive amplitude.

The model uses the virtual nucleon approximation (VNA) [11, 12, 13] to describe DIS from the deuteron. In the VNA, only the proton-neutron component of the deuteron system is considered. It is also assumed that the negative energy projection of the virtual nucleon propagator gives negligible contributions to the scattering amplitude and the contribution of meson exchange currents is neglected. In the VNA, the nuclear wave function $\Psi_D(p)$ is normalized to account for baryon number conservation [14]:

$$\int \alpha |\Psi_D(p)|^2 d^3p = 1, \tag{1}$$

where $\alpha = 2 - \frac{2(E_s - p_{s,z})}{M_D}$ is the light cone momentum fraction of the deuteron carried by the bound nucleon, with $E_s(p_s)$ the on-shell energy (momentum) of the spectator nucleon. Because of the virtuality of the interacting nucleon, it is impossible to satisfy the momentum sum rule at the same time, which can be qualitatively interpreted as part of the deuteron momentum fraction being distributed to non-nucleonic degrees of freedom.

The final-state interaction contribution is encoded in a Feynman diagram where after the interaction of the virtual photon with the bound nucleon, the produced hadronic mass X interacts with the spectator nucleon through an effective rescattering amplitude. At the energies for the experiments considered here, we can assume that the rescattering will be highly diffractive and will occur over small angles and the generalized eikonal approximation (GEA) [15] can be applied. In the GEA an eikonal form is adopted for the effective rescattering amplitude:

$$f_{X'N,XN} = \sigma(Q^2, x)(i + \epsilon(Q^2, x))e^{\frac{B(Q^2,x)}{2}t}, \tag{2}$$

where σ, ϵ and B are the effective total cross section, real part and the slope factor of the diffractive $X'N \to XN$ scattering amplitude. The x and Q^2 dependence of the scattering parameters σ and B were determined by fitting our model calculations to data for tagged spectator deuteron DIS taken in the Deeps experiment at Jefferson Lab [9, 16]. In the derivation of the FSI amplitude, a factorized approach is used, whereby the interaction of the virtual photon with the off-shell nucleon (encoded in the nucleon structure functions) is taken out of the integration over the intermediate spectator momentum. Finally, the four structure functions for tagged spectator deuteron DIS (F_T, F_L, F_{LT}, F_{TT}) can be written as the product of a distorted deuteron momentum distribution containing the amplitude of Eq. (2), the nucleon structure function F_2^N and kinematical factors (for detailed expressions see Ref. [9]).

By applying the optical theorem to the forward virtual Compton scattering amplitude, the model was extended to inclusive deuteron DIS in Ref. [8] :

$$W_D^{\mu\nu} = \frac{1}{2\pi M_D}\frac{1}{3}\sum_{s_D} \Im m\, \mathcal{A}^{\mu\nu}_{\gamma^* D}(t=0)\,, \tag{3}$$

where $W_D^{\mu\nu}$ is the hadronic tensor of the inclusive DIS process and $\mathcal{A}^{\mu\nu}_{\gamma^* D}$ the virtual Compton scattering amplitude. The advantage of such an approach is that the amplitudes accounting for the FSI effects will self-consistently satisfy the unitarity conditions for inelastic rescattering. The amplitude $\mathcal{A}^{\mu\nu}_{\gamma^* D}$ contains contributions from the plane-wave diagram and the rescattering diagram, where the produced hadronic mass X interacts with the spectator nucleon. Again, a factorized approximation is applied and for the intermediate hadronic states three effective resonances (at invariant mass $W = 1.232, 1.5$, and 1.75 GeV) were taken into account in addition to a broad contribution from the DIS continuum. We showed that in our formalism, FSI effects naturally disappear as Q^2 increases due to phase-space constraints and found sizeable FSI effects for $x > 0.6$ and $Q^2 < 10$ GeV^2.

It is now straightforward to extend the model to the situation of a polarized deuteron target. It suffices to replace the average over the deuteron polarization by a trace with the appropriate spin 1 density matrix. Consequently, Eq. (3) can be written as

$$W_D^{\mu\nu} = \frac{1}{2\pi M_D} \sum_{\lambda\lambda'} \Im m\, \rho^D_{\lambda'\lambda} \left[\mathcal{A}^{\mu\nu}_{\gamma^* D}(t=0)\right]_{\lambda\lambda'}, \tag{4}$$

where ρ^D represents the spin 1 density matrix for the deuteron.

3. Deuteron density matrix

The density matrix for a spin 1 particle in a rest frame for the particle can be written in a multipole expansion [17]

$$\rho^D_{\lambda\lambda'} = \frac{1}{3} \sum_{L=0}^{2} \sum_{M=-L}^{L} (2L+1) t^*_{LM} \left(\hat{T}^L_M\right)_{\lambda\lambda'}, \tag{5}$$

where λ denotes the spin projection of the deuteron, t_{LM} are the multipole parameters, and the spherical tensor operators read $\left(\hat{T}^L_M\right)_{\lambda\lambda'} = \langle s\lambda | s\lambda'; LM\rangle$. When the ensemble is made up of particles quantized along the z-axis, the density matrix takes on a simple form

$$\rho^D = \frac{1}{3}\mathrm{diag}(1 + \frac{3}{2}P_z + 2P_{zz}, 1 - P_{zz}, 1 - \frac{3}{2}P_z + 2P_{zz}), \tag{6}$$

where $P_z = \sqrt{2}t_{10} = p^+ - p^-$ and $P_{zz} = \sqrt{10}t_{20} = p^+ + p^- - 2p^0$, with p^λ the probability of finding the deuteron with a spin projection λ along the x-axis.

In theoretical calculations for deuteron DIS, the hadronic tensor is usually evaluated in the hadron plane (determined by the virtual photon and spectator nucleon), with the z-axis along the virtual photon momentum. For a fixed target experiment, however, the deuteron density matrix is usually determined in a deuteron rest frame with the z-axis along the incoming beam. Consequently, care has to be taken to transform the density matrix to the hadron plane in order to do meaningful comparisons with experimental data. This involves a rotation over the angle between the incoming beam and virtual photon θ_{eq}, and over the angle ϕ between the electron and hadron planes. Due to the simple transformation properties under rotations of the multipole parameters in Eq. (5), this transformation can be readily computed [17]. For an ensemble with only the multipole parameter t_{20} nonzero in the rest frame with the z-axis along the incoming beam, one obtains for the multipole parameters $\hat{t}_{LM}$ in the hadron plane [18]:

$$\hat{t}_{20} = \frac{1}{4}(1 + 3\cos 2\theta_{eq})\, t_{20}\,,$$
$$\hat{t}_{21} = -\sqrt{\frac{3}{8}}\sin 2\theta_{eq} e^{i\phi}\, t_{20}\,,$$
$$\hat{t}_{22} = \sqrt{\frac{3}{32}}(1 - \cos 2\theta_{eq})e^{i2\phi}\, t_{20}\,. \quad (7)$$

For the deuteron DIS formalism under consideration here, this means that even after integrating over the angle ϕ (which is done in the inclusive DIS formalism), the cross section will receive contributions from the F_{LT} and F_{TT} response functions (that come with a $\cos\phi$ and $\cos 2\phi$ dependence) through respectively the $\hat{t}_{21}$ and $\hat{t}_{22}$ parts of the density matrix.

4. Results

The cross section for DIS from a polarized deuteron target without any beam polarization can be written as

$$d\sigma = d\sigma_u(1 + \frac{1}{2}P_{zz}A_{zz})\,, \quad (8)$$

where σ_u is the cross section for an unpolarized target, P_{zz} was defined in Eq. (6), and A_{zz} is the tensor asymmetry of the deuteron cross section which can be related to the structure function b_1 [6]. For the calculations of A_{zz} for inclusive DIS from the deuteron, we include resonances at invariant mass $W = 1.232, 1.5$, and 1.75 GeV as the set of effective hadron states entering the final-state interaction amplitude. The values of the scattering parameters in Eq. (2) are taken from our fits to the JLab Deeps data. We used the SLAC nucleon structure function parametrization [19], and for the non-relativistic deuteron wave function we use the parametrization based on the Paris NN potential [20].

In Fig. 1, we present calculations of A_{zz} for inclusive DIS from the deuteron for kinematics accessible in the upcoming JLab experiment [7]. For the plane-wave calculations, we observe that A_{zz} is almost zero for Bjorken x in the DIS region, while it reaches values of the order of $\sim \pm 0.01$ in the resonance region at $Q^2 = 2$ GeV^2 and becomes smaller with increasing Q^2. Adding the FSI contributions has a significant effect on the size of A_{zz} over the whole x range. At $Q^2 = 2$ GeV^2 the off-shell contribution of the FSI diagram also has a sizeable contribution while at $Q^2 = 5$ GeV^2 it is almost negligible. It is worth remarking that even though the effective hadronic states taken into account in the FSI diagram all lie in the resonance region, they also contribute significantly to A_{zz} in the DIS region. This can be understood in the following manner. In the formalism A_{zz} is only nonzero because of the D-wave component of the deuteron wave function, both in the plane-wave and FSI contribution to the cross section. As the dominant contribution for the D-wave occurs at momenta above 250 MeV, A_{zz} is still sensitive to the resonance region FSI contributions in the DIS x-region for free nucleon kinematics due to the Fermi motion of the struck nucleon. Comparing the effect of tensor polarization along the virtual photon with polarization along the incoming beam, we find limited differences for $Q^2 = 5$ GeV^2. For $Q^2 = 2$ GeV^2 the size of A_{zz} decreases, especially at the largest x values, where the angle θ_{eq} is largest.

In Fig. 2, we show similar A_{zz} calculations as in Fig. 1 but for kinematics covered in the Hermes experiment [6]. Similar observations as in Fig. 1 apply. The difference between the left and right panel for $Q^2 = 1$ GeV^2 is more pronounced here because of larger θ_{eq} values, with A_{zz} values including FSIs becoming significantly smaller for the case of polarization along the incoming beam. Comparing our calculation to the Hermes data point of $A_{zz} = 0.157 \pm 0.69$ at $x = 0.45$, $Q^2 \approx 5$GeV^2, we find a value of $A_{zz} \approx 0.0015$, about two orders of magnitude smaller.

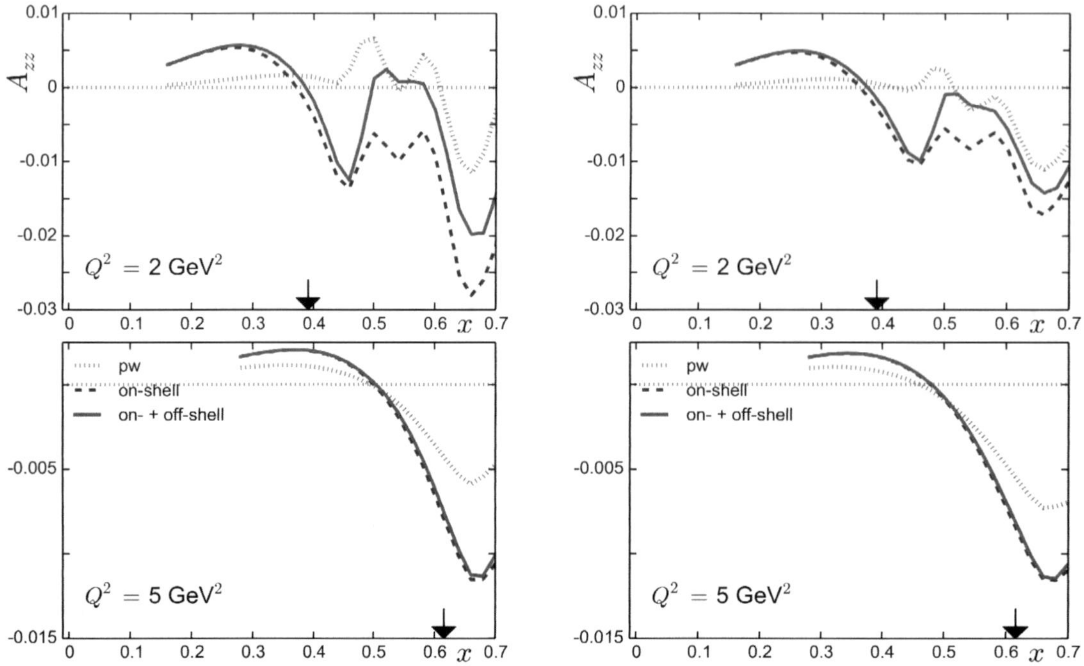

Figure 1. Tensor asymmetry A_{zz} for kinematics accessible in the upcoming JLab experiment [7] (incoming beam of 11 GeV). The red dotted curve shows the plane-wave calculations, the blue dashed (green full) curve includes the contribution of the on-shell (and off-shell) FSI contribution. For details on the FSI on- and off-shell calculations, we refer to Ref. [8]. Left panels have the deuteron polarized along the virtual photon direction, right panels along the incoming beam, with the transformation according to Eq. (7). The arrow along the x-axis indicates the boundary at $W = 2$ GeV between the resonance and DIS regions for free nucleon kinematics.

Finally, in Fig. 3, we show calculations for tagged spectator DIS $[D(e, e'p_s)X]$ from the deuteron at different spectator momenta and two values of the invariant mass W of the produced X, for a deuteron polarized along the virtual photon momentum. One observes that the plane-wave calculations produce sizeable values for A_{zz} in the case of intermediate spectator momenta of 340 MeV (where the deuteron D-wave contribution is largest), especially for forward spectator angles. Including FSIs has little effect at the low spectator momentum of 100 MeV, but causes significant changes for higher spectator momenta in the forward region, where we know from our previous study [9] that FSI effects are largest in the unpolarized cross section. For $p_s = 340$ MeV and $\cos\theta_s > 0.5$ including the FSI contribution even reverses the sign of A_{zz}.

5. Conclusion

We presented calculations for the deuteron tensor asymmetry A_{zz} in a model based on the virtual nucleon and generalized eikonal approximations that allows for the description of final-state interaction effects for deuteron DIS reactions from intermediate hadronic states in the resonance region. For inclusive DIS, we included resonances at invariant mass $W = 1.232, 1.5$, and 1.75 GeV as the set of effective hadron states entering the final-state interaction amplitude. Our calculations for kinematics accessible at JLab12 and Hermes showed sizeable FSI effects both in the resonance and DIS region for free-nucleon kinematics, though the overall size of

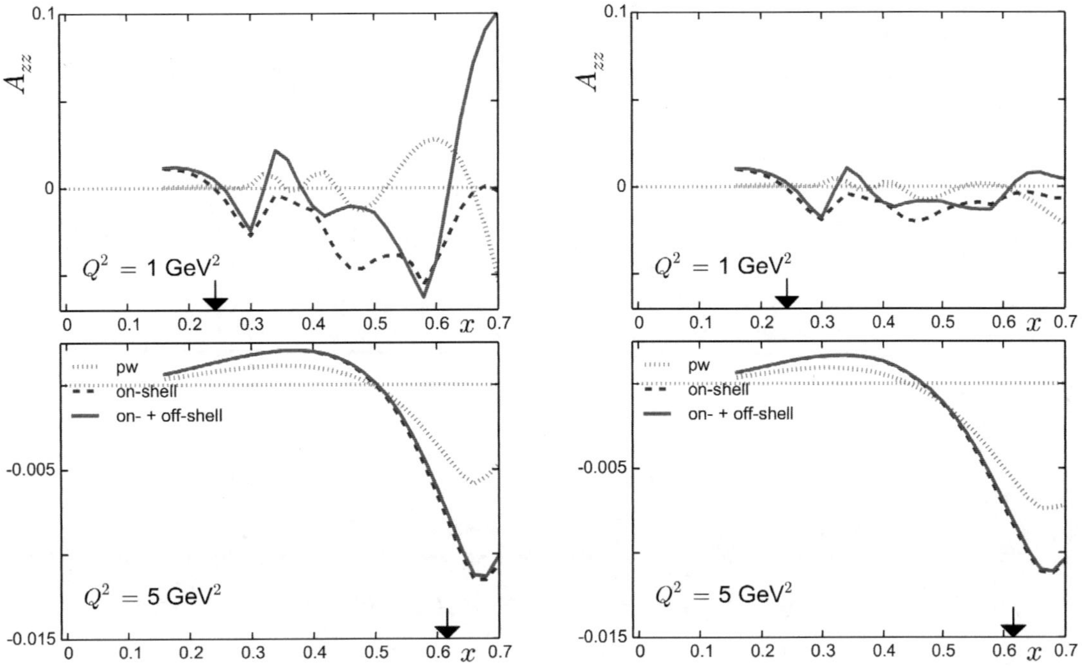

Figure 2. Tensor asymmetry A_{zz} for kinematics covered in the Hermes experiment [6] (incoming beam of 27.6 GeV). Curves and panels as in Fig. 2

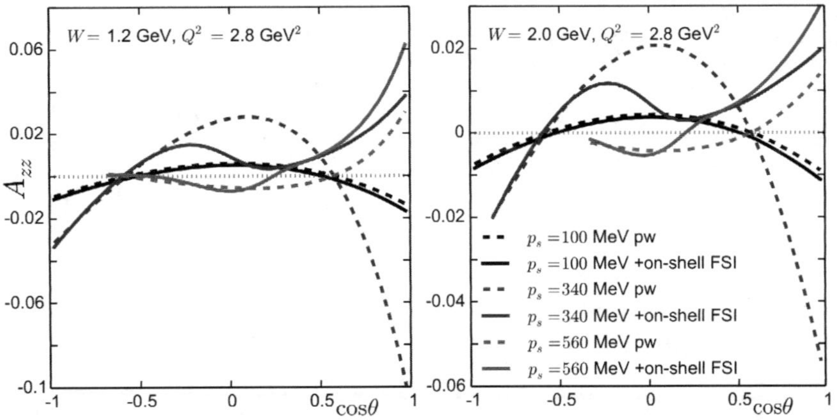

Figure 3. Tensor asymmetry for tagged spectator DIS $D(e, e'p_s)X$ at invariant mass W of X 1.2 (left panel) and 2.0 GeV (right panel) for three different spectator momenta. Full (dashed) curves represent a plane-wave (including FSIs) calculation. For details on the FSI calculations, see Ref. [9].

the nucleonic contribution to A_{zz} remains small. The sensitivity of A_{zz} in the DIS region to FSI contributions from the resonance region can be understood from the dependence of the

A_{zz} observable on the D-wave part of the deuteron wave function. We showed that deuteron polarization along the incoming beam reduces the size of A_{zz} at the low Q^2 values compared to polarization along the virtual photon direction. For the tagged spectator DIS process FSI effects proved to be largest at spectator momenta around 300 MeV and for spectator angles $\cos\theta_s > 0.5$.

Acknowledgments

The computational resources (Stevin Supercomputer Infrastructure) and services used in this work were provided by Ghent University, the Hercules Foundation and the Flemish Government department EWI. This work is supported by the Research Foundation Flanders as well as by the U.S. Department of Energy Grant under Contract DE-FG02-01ER41172.

References

[1] Hoodbhoy P, Jaffe R and Manohar A 1989 *Nucl.Phys.* **B312** 571
[2] Nikolaev N N and Schafer W 1997 *Phys.Lett.* **B398** 245–251 (*Preprint* `hep-ph/9611460`)
[3] Edelmann J, Piller G and Weise W 1997 *Z.Phys.* **A357** 129–131 (*Preprint* `nucl-th/9701026`)
[4] Bora K and Jaffe R 1998 *Phys.Rev.* **D57** 6906–6911 (*Preprint* `hep-ph/9711323`)
[5] Miller G A 2014 *Phys.Rev.* **C89** 045203 (*Preprint* `1311.4561`)
[6] Airapetian A *et al.* (HERMES Collaboration) 2005 *Phys.Rev.Lett.* **95** 242001 (*Preprint* `hep-ex/0506018`)
[7] Slifer K and Long E 2014 *PoS* **PSTP2013** 008 (*Preprint* `1311.4835`)
[8] Cosyn W, Melnitchouk W and Sargsian M 2014 *Phys.Rev.* **C89** 014612 (*Preprint* `1311.3550`)
[9] Cosyn W and Sargsian M 2011 *Phys. Rev.* **C84** 014601 (*Preprint* `1012.0293`)
[10] Cosyn W and Sargsian M 2011 *AIP Conf.Proc.* **1369** 121–128 (*Preprint* `1101.1258`)
[11] Melnitchouk W, Sargsian M and Strikman M I 1997 *Z. Phys.* **A359** 99–109 (*Preprint* `nucl-th/9609048`)
[12] Sargsian M and Strikman M 2006 *Phys. Lett.* **B639** 223–231 (*Preprint* `hep-ph/0511054`)
[13] Sargsian M M, Simula S and Strikman M I 2002 *Phys. Rev.* **C66** 024001 (*Preprint* `nucl-th/0105052`)
[14] Frankfurt L L and Strikman M I 1981 *Phys. Rept.* **76** 215–347
[15] Sargsian M M 2001 *Int. J. Mod. Phys.* **E10** 405–458 (*Preprint* `nucl-th/0110053`)
[16] Klimenko A V *et al.* (CLAS) 2006 *Phys. Rev.* **C73** 035212 (*Preprint* `nucl-ex/0510032`)
[17] Leader E 2005 *Spin in Particle Physics (Cambridge Monographs on Particle Physics, Nuclear Physics and Cosmology)* (Cambridge University Press) ISBN 0521020778
[18] Jeschonnek S and Van Orden J W 2009 *Phys.Rev.* **C80** 054001 (*Preprint* `0907.3712`)
[19] Bodek A, Breidenbach M, Dubin D L, Elias J E, Friedman J I, Kendall H W, Poucher J S, Riordan E M, Sogard M R, Coward D H and Sherden D J 1979 *Phys. Rev. D* **20** 1471–1552
[20] Lacombe M *et al.* 1980 *Phys. Rev.* **C21** 861–873

Journal of Physics: Conference Series **543** (2014) 012007 doi:10.1088/1742-6596/543/1/012007

Neutron spin structure with polarized deuterons and spectator proton tagging at EIC

W Cosyn[1], V Guzey[2], D W Higinbotham[3], C Hyde[4], S Kuhn[4], P Nadel-Turonski[3], K Park[4], M Sargsian[5], M Strikman[6], C Weiss[3]

[1] Ghent University, 9000 Gent, Belgium
[2] Petersburg Nuclear Physics Institute, Gatchina, 188300, Russia
[3] Jefferson Lab, Newport News, VA 23606, USA
[4] Old Dominion University, Norfolk, VA 23529, USA
[5] Florida International University, Miami, FL 33199, USA
[6] Pennsylvania State University, University Park, PA 16802, USA

E-mail: weiss@jlab.org

Abstract. The neutron's deep–inelastic structure functions provide essential information for the flavor separation of the nucleon parton densities, the nucleon spin decomposition, and precision studies of QCD phenomena in the flavor–singlet and nonsinglet sectors. Traditional inclusive measurements on nuclear targets are limited by dilution from scattering on protons, Fermi motion and binding effects, final–state interactions, and nuclear shadowing at $x \ll 0.1$. An Electron–Ion Collider (EIC) would enable next–generation measurements of neutron structure with polarized deuteron beams and detection of forward–moving spectator protons over a wide range of recoil momenta ($0 < p_R <$ several 100 MeV in the nucleus rest frame). The free neutron structure functions could be obtained by extrapolating the measured recoil momentum distributions to the on-shell point. The method eliminates nuclear modifications and can be applied to polarized scattering, as well as to semi-inclusive and exclusive final states. We review the prospects for neutron structure measurements with spectator tagging at EIC, the status of R&D efforts, and the accelerator and detector requirements.

1. Introduction

The program of exploring short-range nucleon structure and strong interaction dynamics with high–energy lepton scattering relies on measurements on the neutron as much as those on the proton target. Neutron and proton data are needed to separate the isoscalar and isovector combinations of the deep-inelastic scattering (DIS) structure functions, which are subject to different short–distance dynamics (QCD evolution, higher–twist effects, small–x behavior) and give access to different combinations of the parton densities (gluons and singlet quarks vs. non-singlet quarks). The unpolarized isovector structure function $F_{2p} - F_{2n}$ constrains the flavor composition of the nucleon's sea quark densities and permits study of small–x dynamics in the non-singlet sector. In the polarized case the isovector structure function $g_{1p} - g_{1n}$ can be used to cleanly separate leading–twist and higher–twist dynamics, taking advantage of its simple QCD evolution (no mixing with gluons). The isoscalar combination $g_{1p} + g_{1n}$ can then be used to extract the polarized gluon density from the leading–twist QCD evolution. Both isospin combinations are needed to determine the flavor decomposition of the polarized quark densities and their contributions to the nucleon spin. Neutron and proton data combined are needed also

Journal of Physics: Conference Series **543** (2014) 012007 doi:10.1088/1742-6596/543/1/012007

to investigate the dynamical mechanism causing single–spin asymmetries in semi-inclusive DIS (where there are hints of surprisingly large isovector structures) and to constrain the generalized parton distributions in deeply–virtual Compton scattering.

Neutron structure is measured in high–energy scattering experiments with nuclear targets. The extraction of the free neutron structure functions from nuclear DIS data faces considerable challenges (see Ref. [1] for a review). Nuclear binding modifies the apparent neutron structure functions through the Fermi motion and other dynamical effects. At moderate momentum transfers $Q^2 \sim$ few GeV^2 the nuclear cross sections are affected by final–state interactions involving other nucleons. At $x \ll 0.1$ the large coherence length of the electromagnetic probe causes quantum–mechanical interference of the amplitudes for scattering from different nucleons along its path, which results in shadowing and antishadowing effects in the cross section [2]. In polarized measurements with light nuclei (deuteron $D \equiv {}^2$H, ^{3}He) there is significant dilution from scattering on the proton(s). One must know not only the degree of neutron polarization in the nucleus but also the spin dependence of the various nuclear modifications of nucleon structure. Attempts to account for these nuclear effects theoretically are complicated by the fact that they arise as an average over different classes of configurations in the nuclear wave function, which cannot be separated in inclusive measurements. It is clear that better experimental control of the nuclear environment is needed to improve the precision of neutron structure extraction. This is particularly relevant if a combination of neutron and proton data are to be used to study subtle QCD effects, such as the separation of leading and higher twist, non-singlet QCD evolution, and non-singlet small–x behavior.

Inclusive deep–inelastic scattering from nuclei was measured in fixed–target experiments with electron and muon beams, at electron–nucleon squared center–of–mass energies $s_{eN} \equiv s_{eA}/A$ in the range $s_{eN} \sim 10 - 900\,\text{GeV}^2$ (JLab 6 GeV, DESY HERMES, SLAC, CERN EMC/NMC, FNAL E665; see Refs. [1, 3] for a review of the data). Measurements of neutron spin structure with polarized nuclear targets were performed at SLAC, DESY HERMES, CERN SMC/COMPASS and JLab 6 GeV [4, 5, 6]. Both unpolarized and polarized nuclear measurements will be extended with the JLab 12 GeV Upgrade [3]. While these experiments have provided basic information on neutron structure at $x \gtrsim 0.01$, addressing the above questions will require data of much higher precision and wider kinematic coverage.

The Electron–Ion Collider (EIC), proposed as a next-generation facility for nuclear physics, would dramatically expand the opportunities for high-energy scattering on polarized light nuclei (D, ^{3}He, ...) and measurements of neutron structure.[1] The medium–energy EIC designs recently developed provide electron–nucleon squared center–of–mass energies in the range $s_{eN} \sim 250 - 2500\,\text{GeV}^2$ at luminosities up to $\sim 10^{34}\,\text{cm}^{-2}\,\text{s}^{-1}$ [8]. The wide kinematic coverage would permit definitive studies of the Q^2 evolution and small–x behavior of structure functions; the high luminosity would enable measurements of spin asymmetries and rare processes involving exceptional configurations. Even more important, two specific capabilities provided by the EIC would allow one to control the nuclear modifications and extract neutron structure with unprecedented precision.

One capability are deuteron beams, especially polarized deuterons, as would be available for the first time with the JLab MEIC, thanks to the figure–8 shape of the ion ring designed to compensate the effect of spin precession [9]. The deuteron is the simplest nucleus ($A = 2$); its wave function is known well up to large nucleon momenta ~ 300 MeV, including the light-front wave function describing microscopic nuclear structure as probed in high-energy scattering processes [10]. The deuteron has spin 1 and is mostly in the $L = 0$ configuration (S–wave), with a small admixture of $L = 2$ (D–wave), such that the proton and neutron are spin–polarized and their degree of polarization is known very well. Because there are only two nucleons the

[1] For a general overview of the medium–energy EIC physics program, including measurements with proton beams, see e.g. Ref. [7].

possibilities for final–state interactions are limited; in the configurations where they can happen they can be estimated using theoretical models (see e.g. Refs. [11, 12] for a model in the resonance region $W \sim$ few GeV).

The other capability is the detection of spectator nucleons emerging from the high–energy scattering process ("spectator tagging"). In collider experiments the spectator nucleons carry a fraction $\sim 1/A$ of the ion beam momentum and can be detected with appropriate forward detectors. The technique is uniquely suited to colliders: there is no target material absorbing low-momentum nucleons, and it can be used with polarized ion beams (longitudinal and transverse). Spectator tagging is especially powerful in scattering on the deuteron. It allows one to positively identify the active neutron through proton tagging, and to control its quantum state through measurement of the recoil momentum. Spectator proton tagging with unpolarized deuterium was explored in a pioneering fixed-target experiment at JLab with 6 GeV beam energy (CLAS BoNuS detector, covers recoil momenta $p_R \gtrsim 70\,\text{MeV}$) [13] and will be studied further at 11 GeV.

In this note we summarize the potential of DIS on the deuteron with spectator proton tagging for precision measurements of the neutron structure functions at EIC. We describe the method for eliminating nuclear structure effects through on-shell extrapolation in the spectator proton momentum, present simulations of unpolarized and polarized observables [14], and outline the accelerator and detector requirements. It should be noted that the various nuclear modifications of the nucleon's partonic structure are interesting physics topics in their own right (with connections to short-range NN correlations, non-nucleonic degrees of freedom, etc.) and can be studied with the same spectator tagging measurements at larger recoil momenta [15].

2. Neutron structure with spectator tagging

The basic method for extracting the free neutron structure function with spectator tagging is described in Ref. [16] (see Fig. 1). One measures the cross section of conditional DIS $e + D \rightarrow e' + p + X$ as a function of the recoil proton momentum, parametrized by the light–cone fraction $\alpha_R \equiv 2(E_R + p_R^z)/(E_D + p_D^z)$ and the transverse momentum $\boldsymbol{p}_{RT}$, defined in a frame where the deuteron momentum $\boldsymbol{p}_D$ and the $\boldsymbol{q}$ vector are collinear and along the z–direction (see Fig. 1a). A key variable is the invariant 4–momentum transfer between the deuteron and the recoil proton, $t \equiv (p_R - p_D)^2$, calculated from α_R and $\boldsymbol{p}_{RT}$. As a function of t the scattering amplitude has a pole at $t = M_N^2$, which arises from the impulse approximation diagram of Fig. 1b and corresponds to "neutron exchange" in the t–channel. The residue at the pole is, up to a constant factor representing deuteron structure, given by the structure function of the *free neutron,* evaluated at the argument $\tilde{x} = x/(2 - \alpha_R)$, where $x = Q^2/(p_D q)$ is the scaling variable with $0 < x < 2$. [2] It can be shown that nuclear binding and final–state interactions only affect the amplitude at $M_N^2 - t > 0$, but not the residue at the pole [16].

To extract the free neutron structure function one plots the measured tagged cross section as a function of t, removes the pole factor $1/(M_N^2 - t)^2$, and extrapolates to $t \rightarrow M_N^2$ [16]. The pole in t is extremely close to the physical region, so that the extrapolation can be performed with great accuracy; the minimum physical value of $M_N^2 - t$ is $\epsilon_D M_D$, where ϵ_D is the deuteron binding energy. The method is analogous to the Chew–Low extrapolation used to extract pion structure from πN scattering data. Figure 1c shows a simulated on-shell extrapolation with MEIC pseudodata ($s_{eN} = 1000\,\text{GeV}^2$, integrated luminosity $10^6\,\text{nb}^{-1}$) [14]. One sees that the t–dependence is very smooth. The extrapolation in t is performed at *fixed* values of the recoil light-cone fraction α_R, such that the effective $\tilde{x}$ in the neutron structure function remains constant along the way. Figure 1c shows the pseudodata in three bins of α_R with average values $\langle \alpha_R \rangle = (1, 1.04, 1.08)$ and a width of 0.04. Note that the kinematic limit in $M_N^2 - t$ increases

[2] The variables are defined such that in the absence of nuclear binding $\alpha_R = 1$, and $x = \tilde{x}$ coincides with the usual scaling variable for scattering from a free nucleon.

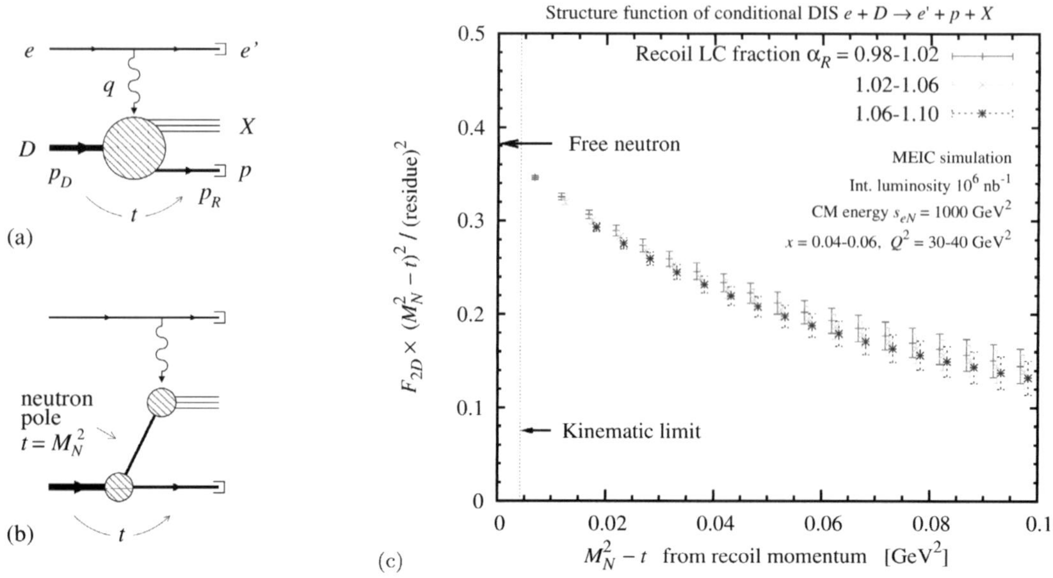

Figure 1. (a) Conditional DIS on the deuteron, $e + D \to e' + p + X$. (b) Neutron pole at $t = M_N^2$ arising from the impulse approximation diagram. (c) Simulated measurement of the recoil momentum dependence with MEIC and on-shell extrapolation $t \to M_N^2$ [14]. The plot shows the conditional structure function, with the pole factor $1/(M_N^2 - t)^2$ and the residue removed, as a function of the off-shellness $M_N^2 - t$ calculated from the measured proton recoil momentum. The error bars indicate the expected statistical errors. The three sets correspond to measurements in different intervals of the recoil light–cone momentum fraction α_R.

from $\epsilon_D M_D$ as α_R moves away from unity. Comparison of the extrapolation results at different α_R allows one to confirm the universality of the nucleon pole and provided an important test of the theoretical framework. Critical to the success of the method is the ability to measure the recoil proton momentum with complete coverage down to $p_{RT} = 0$, and with a resolution $\Delta p_{\mathrm{RT}} \lesssim 20\,\mathrm{MeV}$ and $\Delta\alpha_R \lesssim 10^{-3}$.

The method described here permits a clean and model-independent extraction of free neutron structure and would enable precision measurements of F_{2n} over the entire x–Q^2 range accessible with EIC. Combined with proton data taken in the same kinematics it would determine the isovector structure function $F_{2p} - F_{2n}$, which constrains the flavor structure of the nucleon sea. The results could be cross–checked against those from dilepton production in $pp/\bar{p}p$ collisions, which separate quark flavors through selection of γ^* (Drell–Yan), Z, and $W^\pm$ events, and be used to test the universality of the sea quark densities [17]. The isovector $F_{2p} - F_{2n}$ would also permit quantitative studies of small–x dynamics in the non-singlet sector, which is largely unexplored and raises many interesting questions (QCD structure of Reggeon exchange, non-singlet diffraction, etc.). In practice such studies will likely be limited to the region $x > 0.01$, as the difference between proton and neutron structure functions becomes very small at low x [$(F_{2p} - F_{2n})/(F_{2p} + F_{2n}) < 0.02$ at $x < 0.01$ for $Q^2 \sim$ several 10 GeV^2] and would be increasingly difficult to determine from separate measurements. Lastly, the results for F_{2n} itself would serve as stringent test for models of nuclear effects in heavier nuclei.

DIS on the deuteron with spectator proton tagging could be used to address other interesting physics questions besides free neutron structure. The recoil momentum dependence of the effective neutron structure function in $e + D \to e' + p + X$ allows one to study the modification of the neutron's partonic structure as a function of its momentum in the deuteron wave function — a qualitative improvement compared to usual inclusive measurements. Theoretical arguments suggest that the modification of the effective neutron structure function in the deuteron is in first approximation proportional to the neutron's virtuality $M_N^2 - t \approx 2|\boldsymbol{p}_R|^2$(rest frame), which is controlled by the measured recoil momentum [18]. Furthermore, the modification is related in a simple way to the EMC effect in inclusive nuclear structure functions of heavier nuclei ($A > 2$) and exhibits similar x–dependence. Measurements of the recoil momentum dependence of the effective neutron structure function in $e + D \to e' + p + X$ thus provide direct insight into the dynamical origin of the EMC effect. In particular, measurements at large recoil momenta $|\boldsymbol{p}_R|$(rest frame) $\gg 100\,\text{MeV}$ could verify a possible connection between the EMC effect and short–range NN correlations in nuclei [19, 20].

Measurements of the recoil momentum dependence in $e + D \to e' + p + X$ at finite $|\boldsymbol{p}_{RT}|$ and $\alpha_R \neq 1$ also provide a unique method for studying nuclear final–state interactions in DIS at $Q^2 \sim$ few GeV^2. The results of such measurements could be used to refine phenomenological models of the final–state interactions [11, 12], which would improve the precision of neutron structure extraction from inclusive nuclear scattering data (including $A > 2$).

3. Neutron spin structure

Polarized DIS on the deuteron with spectator proton tagging can be used to determine the free neutron spin structure function g_{1n}. The method for eliminating nuclear effects described above can straightforwardly be extended to double–polarized scattering. The simplest case is that of a longitudinally polarized electron beam colliding with a longitudinally polarized deuteron beam, where "longitudinal" refers to the respective beam directions.[3] One measures the cross section of conditional double–polarized DIS $\vec{e} + \vec{D} \to e' + p + X$ as a function of the recoil proton momentum variables; the recoil proton polarization remains undetected and is summed over. The on-shell extrapolation in t could in principle be performed directly with the polarized cross sections, using the same formulas as in the unpolarized case. It is more convenient, however, to work with the spin asymmetry

$$A_{\parallel} = \frac{\sigma(++) - \sigma(+-)}{\sigma(++) + \sigma(+-)}, \tag{1}$$

where $\sigma(\lambda_e, \lambda_D)$ denotes the conditional cross section for the spin projections $\lambda_e = \pm 1/2$ and $\lambda_D = \pm 1$. In Eq. (1) the nucleon pole $1/(M_N^2 - t)^2$ in the cross sections cancels between numerator and denominator (as do most of the nuclear modifications at $M_N^2 - t > 0$), and the asymmetry naturally has a smooth dependence on t. Figure 2 shows a simulated measurement of the conditional spin asymmetry at MEIC as a function of t ($s_{eN} = 1000\,\text{GeV}^2$, integrated luminosity $2 \times 10^7\,\text{nb}^{-1}$ = 20 times higher than in the unpolarized simulation of Fig. 1c) [14]. The projections shown here were made using a simple model of deuteron structure that does not include the D–wave, assumes the same nuclear modification of polarized and unpolarized cross sections at $M_N^2 - t > 0$, and therefore does not have an explicit t–dependence of the asymmetry; in a more realistic model there would be a weak residual t–dependence at $M_N^2 - t > 0$ due to the said effects. Figure 2 shows that an accurate measurement of the conditional spin asymmetry could be made over a range of $M_N^2 - t \lesssim 0.05\,\text{GeV}^2$ with the proposed setup, and that the on-shell extrapolation could be performed as in the unpolarized case. More elaborate simulations, assuming general deuteron polarization (mixed states, tensor polarization) and using realistic deuteron wave functions, are in progress [14].

[3] In the present MEIC design the electron and deuteron beam collide with a finite crossing angle $\sim 50\,\text{mrad}$; its effect on the net polarization along the $\boldsymbol{q}$ vector direction can easily be calculated and is very small.

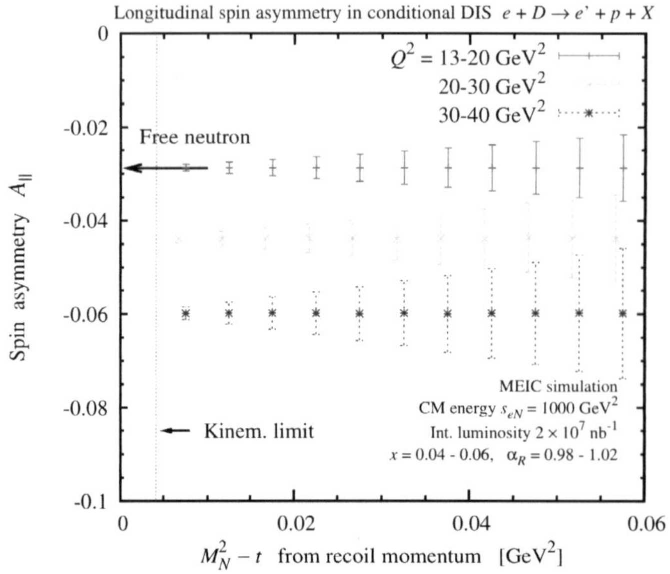

Figure 2. Simulated measurement of the longitudinal double spin asymmetry $A_{||}$, Eq. (1), in conditional DIS on the deuteron with spectator proton tagging, $\vec{e} + \vec{D} \to e' + p + X$ with MEIC. The measured asymmetry is shown as a function of the off–shellness $M_N^2 - t$, calculated from the measured proton recoil momentum, in a single interval of the recoil light-cone momentum fraction α_R. The error bars indicate the expected statistical errors. The three sets correspond to different values of Q^2.

The on-shell (extrapolated) asymmetry can then directly be used to extract the *free neutron* spin structure function. An important simplification occurs because the D–wave contribution to the cross section is very small at the on-shell point, as it is proportional to a higher power of the recoil proton rest–frame momentum. It implies that the deuteron wave function at the pole can be regarded as a practically pure S–wave, in which the neutron is completely polarized along the deuteron spin direction, and renders further analysis as simple as in the case of a polarized nucleon target:

$$A_{||} \text{ (on-shell)} \sim D\, g_{1n}/F_{1n} \qquad \text{(up to terms} \sim M_N^2/Q^2), \tag{2}$$

where D is a kinematic factor (depolarization factor), and the neutron structure functions are evaluated at $\tilde{x} = x/(2 - \alpha_R)$ (complete expressions will be given elsewhere). Spectator tagging with on-shell extrapolation thus overcomes both the dilution and the neutron polarization uncertainties associated with inclusive nuclear measurements.

The method outlined here would enable precision measurements of the neutron spin structure function g_{1n} over a wide kinematic range without any model-dependent nuclear structure input. The neutron data thus obtained could be used in several ways to advance our understanding of nucleon spin structure and QCD. Combined with the proton data they would determine the isovector polarized structure function $g_{1p} - g_{1n}$, which exhibits particularly simple QCD evolution (no mixing with gluons) and can be used to cleanly separate leading–twist and higher–twist dynamics. They would also enable precision tests of the Bjorken sum rule, and possibly a determination of the strong coupling α_s competitive with other methods (see Ref. [21] for a

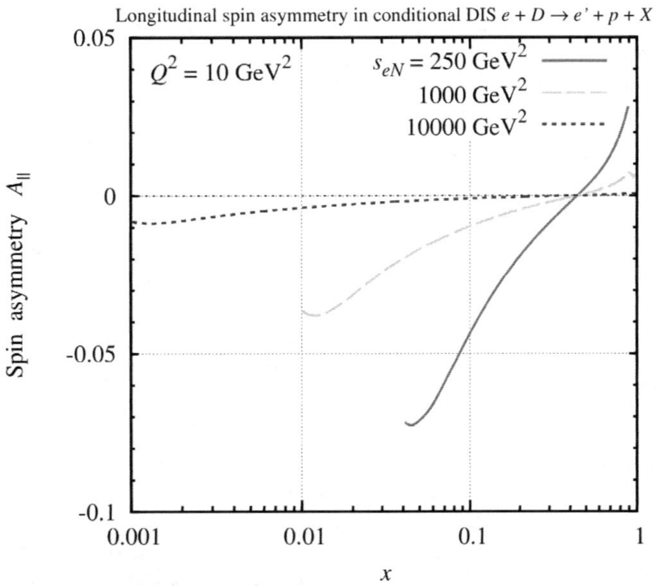

Figure 3. The observable double spin asymmetry $A_{\|}$, Eq. (1), in conditional DIS on the deuteron with spectator proton tagging, $\vec{e}+\vec{D} \rightarrow e'+p+X$, as a function of x, at $Q^2 = 10\,\mathrm{GeV}^2$. The three curves correspond to measurements at squared electron–nucleon center-of-mass energies $s_{eN} = 250, 1000$ and $10000\,\mathrm{GeV}^2$. The kinematic range in x available at a given energy is $x > Q^2/s_{eN}$.

recent analysis using fixed–target data). The isoscalar structure function $g_{1p}+g_{1n}$ could be used to determine the polarized gluon density from the leading–twist QCD evolution. Proton and neutron data combined would allow one to better separate quark and gluon contributions to the nucleon spin, and to determine the flavor composition of the quark spin. The status of global QCD fits to determine the polarized parton densities and the need for EIC data have have been discussed extensively in the literature. A dedicated study of the impact of spectator tagging with polarized deuterons on polarized parton densities is planned.

The center–of–mass energy of the electron–deuteron collision is an important parameter in the spin asymmetry measurements described here. According to Eq. (1) the magnitude of the observable spin asymmetry $A_{\|}$ is determined by the depolarization factor

$$D = \frac{y(2-y)}{2(1-y)(1+R)+y^2} \qquad \text{(up to terms} \sim M_N^2/Q^2), \tag{3}$$

where R is the usual L/T ratio of the unpolarized cross sections, and y is the scaling variable measuring the electron fractional energy loss in the target rest frame. One sees that D decreases proportionally to y if $y \ll 1$. To maximize the observable $A_{\|}$ one wants to keep y at values of order unity. Now at fixed x and Q^2 one has $y \approx Q^2/(xs_{eN})$, where $s_{eN} \equiv s_{eD}/2$. We conclude that one should do the measurement at squared CM energies s_{eN} not much larger than Q^2/x, such that one can keep y of order unity in the region of interest. This is illustrated by Fig. 3, which shows the projected observable spin asymmetry $A_{\|}$ at a fixed Q^2 as a function of x, for different values of s_{eN}. The projected values of the asymmetry here were calculated using the parametrization of the neutron spin structure function provided by the global QCD fit of

Ref. [22] (the differences to more recent fits are unimportant for the point illustrated here). One sees how the observable asymmetry at a given Q^2 and x decreases if measured at "too large" center–of–mass energies. It underscores the need for an EIC design that can deliver high luminosity over a *range of center–of–mass energies* appropriate to the measurements in question.

4. Semi-inclusive and exclusive measurements

The spectator tagging method described here can be extended to measurements in which the final state of the DIS process on the neutron is analyzed further in coincidence with the spectator proton. Measurements of single–inclusive hadron production in the current fragmentation region $e + D \to e' + p + h + X'$ (semi-inclusive DIS on the neutron) could further constrain the partonic structure of the nucleon by tagging the charge and flavor of the struck quark in the neutron. Combined with proton data such measurements could conclusively determine the flavor structure of the light nucleon sea (polarized and unpolarized) and separate strange and non-strange quark densities. Neutron measurements through spectator tagging could also help in exploring the dynamical mechanisms generating single–spin asymmetries in semi-inclusive hadron production (there are hints of large isovector structures, possibly related to dynamical chiral symmetry breaking) and determine the flavor structure of the nucleon's intrinsic transverse momentum distributions. Finally, spectator tagging could be used to measure exclusive processes on the neutron (deeply–virtual Compton scattering, meson production), where the isospin dependence is essential for verifying the reaction mechanism and determining the generalized parton distributions.

5. Accelerator and detector requirements

The spectator tagging measurements described here require integrated forward detectors with complete coverage for protons with low recoil momenta relative to beam momentum per nucleon (p_{RT} < 200 MeV, $p_{R\parallel}/p(\text{beam}) \sim 0.8 - 1.2$), and sufficient recoil momentum resolution ($\Delta p_{RT} \lesssim 20\,\text{MeV}$, $\Delta p_L/p_L \sim 10^{-4}$). The MEIC interaction region and forward detection system have been designed for this purpose and provide fully sufficient capabilities for the physics program outlined here [9]. The physics analysis of spectator tagging data also requires that the intrinsic momentum spread in the ion beam be sufficiently small to allow for accurate reconstruction of the actual recoil momentum at the interaction vertex. Simulations show that with the MEIC beam parameters the "smearing" of the kinematic variables is very moderate and does not substantially affect the physics analysis (the resulting uncertainty in t is of the order $\sim 0.005\,\text{GeV}^2$ — the bin size in Fig. 1c) [14].

The neutron spin structure measurements rely on the polarized deuteron beams that would become available with MEIC. In the figure–8 ring no significant loss of polarization is expected during acceleration and storage of either protons or deuterons, so that the polarization is essentially maintained at the source level. It is expected that a longitudinal vector polarization in excess of 70% could be achieved for a deuteron beam with $\sim$ 50 GeV per nucleon [9], which would be sufficient for the measurements described here.

An R&D program in under way at JLab to develop simulation tools for spectator tagging with EIC (cross section models, event generators) and demonstrate the feasibility of such measurements [14]. The tools are being made available to users and can applied to a variety of processes of interest. Information about available resources may be obtained from the authors.

Acknowledgments

This material is based upon work supported by the U.S. Department of Energy, Office of Science, Office of Nuclear Physics under contract DE-AC05-06OR23177. W Cosyn is supported by Research Foundation Flanders.

Notice: Authored by Jefferson Science Associates, LLC under U.S. DOE Contract No. DE-AC05-06OR23177. The U.S. Government retains a non–exclusive, paid–up, irrevocable, world–wide license to publish or reproduce this manuscript for U.S. Government purposes.

References

[1] Arneodo M 1994 *Phys. Rept.* **240** 301
[2] Frankfurt L, Guzey V and Strikman M 2012 *Phys. Rept.* **512** 255
[3] Malace S, Gaskell D, Higinbotham D W and Cloet I 2014 *Int. J. Mod. Phys. E* **23** 1430013
[4] Kuhn S E, Chen J-P and Leader E 2009 *Prog. Part. Nucl. Phys.* **63** 1
[5] Chen J-P, Deur A, Kuhn S and Meziani Z-E 2011 *J. Phys. Conf. Ser.* **299** 012005
[6] Aidala C A, Bass S D, Hasch D and Mallot G K 2013 *Rev. Mod. Phys.* **85** 655
[7] Accardi A, Guzey V, Prokudin A and Weiss C 2012 *Eur. Phys. J. A* **48** 92
[8] For current information on the EIC machine designs, see: `https://eic.jlab.org/wiki/` (JLab) and `https://wiki.bnl.gov/eic/` (BNL).
[9] Abeyratne S *et al* 2012 *arXiv*:1209.0757
[10] Frankfurt L L and Strikman M I 1981 *Phys. Rept.* **76** 215
[11] Cosyn W, Melnitchouk W and Sargian M 2014 *Phys. Rev. C* **89** 014612
[12] Cosyn W and Sargsian 2011 *Phys. Rev. C* **84** 014601
[13] Tkachenko S *et al* 2014 *Phys. Rev. C* **89** 045206
[14] Weiss C *et al* 2014 Materials of JLab LDRD project "Physics potential of polarized light ions with EIC@JLab," available at: `https://eic.jlab.org/wiki/index.php/Forward_Tagging`
[15] Guzey V, Higinbotham D, Hyde C, Nadel-Turonski P, Park K, Sargsian M, Strikman M, and Weiss C 2014 *arXiv*:1407.3236
[16] Sargsian M and Strikman M 2006 *Phys. Lett. B* **639** 223
[17] Martin A D, Stirling W J, Thorne R S, Watt G 2009 *Eur. Phys. J. C* **63** 189
[18] Frankfurt L L and Strikman M I 1985 *Nucl. Phys. B* **250** 143
[19] Weinstein L B, Piasetzky E, Higinbotham D W, Gomez J, Hen O and Shneor R 2011 *Phys. Rev. Lett.* **106** 052301
[20] Hen O, Higinbotham D W, Miller G A, Piasetzky E and Weinstein L B 2013 *Int. J. Mod. Phys. E* **22**, 1330017
[21] Deur A *et al* 2014, *Phys. Rev. D* **90**, 012009
[22] Gluck M, Reya E, Stratmann M and Vogelsang W 2001 *Phys. Rev. D* **63** 094005

Tensor Polarized Deuteron at an Electron-Ion Collider

N Kalantarians
Hampton University, Hampton, Virginia 23668 USA
E-mail: narbe@jlab.org

Abstract. The deuteron is a simple, spin-1 nuclear system. This makes it an effective test-bed for investigating nuclear physics. In addition, it is also relatively easy to polarize. Tensor polarization provides new opportunities to study properties of the nucleus. This is a significant part of the motivation for a polarized deuteron beam for an electron-ion collider envisioned for the future. This paper will discuss the motivation and a useful physics starting point for such a facility.

1. Motivation

The deuteron is a spin-1 system that can be considered as a simple lab for nuclear physics. Its spatial distribution depends on its spin-state [1]. Tensor polarization presents a unique opportunity to study the spin structure of a nuclear system due to it providing access to the spin ($m = 0, \pm 1$) of the deuteron rather than the 1/2-spin of the neutron or proton only. In essence it is a combination of quark and nuclear physics.

Inclusive scattering incident on a tensor polarized deuteron results in 4 additional spin-structure functions denoted $b_{1,2,3,4}$ [2, 3]. Of these functions, $b_{1,2}$ have been extracted by the HERMES collaboration [4]. The smaller x kinematics of an EIC would also make it possible to access anti-shadowing and coherent (2-body) scattering.

2. A Starting Point: b_1

In essence, the deuteron provides a combination of quark and nuclear physics. Extraction of b_1 is done via deep inelastic scattering (DIS), yet it depends on the deuterons spin-state. This allows for investigation of nuclear effects at the partonic level.

Calculations predict b_1 to vanish in the absence of nuclear effects [2]. Even accounting for the d-state admixture, it is expected to be very small. Various calculations predict b_1 to be on order of $10^{-4} - 10^{-3}$, depending on it taking in to account relativistic convolution with binding [5] or the Bethe-Salpeter formalism [6].

The sole data set with which b_1 was extracted is from HERMES [4]. The observable is tensor asymmetry A_{zz}, which is the normalized cross-section difference. From this b_1 is calculated via the relation $b_1 = 3/2(F_1 A_{zz})$. Here F_1 is the unpolarized structure function. There is a complementary measurement proposed for Jefferson Lab [13].

Predictions for b_1 as $x \to 0$ show it rapidly increasing, indicating double scattering [7, 8]. In addition, predictions for b_2 and A_{zz} indicate possibilities for disentangling predictions at small x [9, 10].

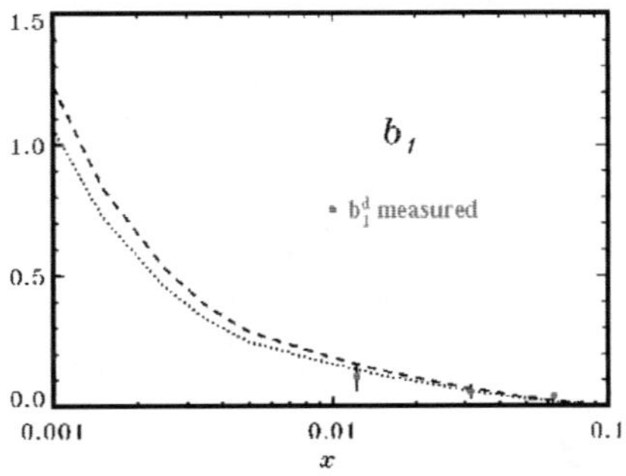

Figure 1. Prediction for b_1 from Ref [9]. Red points are the HERMES results with statistical errors.

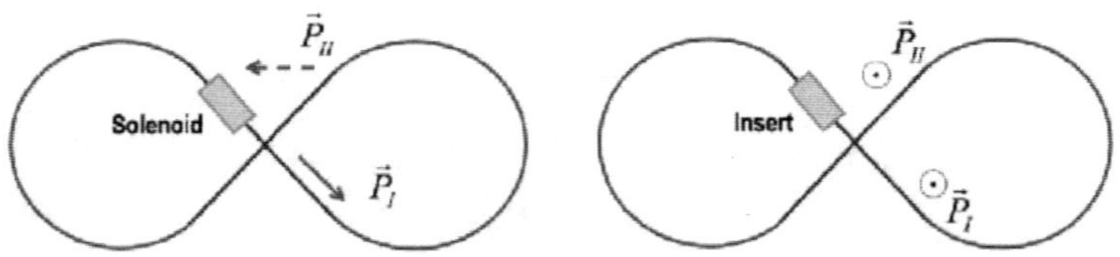

Figure 2. Schematic for polarized deuteron beam for MEIC: longitudinal on left and transverse on right.

3. Studies of Tensor Polarized Deuteron Beam

The staged, medium energy electron-ion collider (MEIC) concept at Jefferson Lab envisions at polarized deuteron beam for various reasons [11]. Due to its small magnetic moment, polarization would be very difficult to maintain for a deuteron beam at sufficiently high energies. The MEIC design calls for an innovative figure-8 ion complex, which would allow the polarization to survive with a spin rotation around a certain axis using magnetic inserts. With this the deuteron polarization would then remain stable and point along the rotation axis at the inserts location. Simulations studies are underway for this.

Deuteron spin manipulation has been studied at IUCF and COSY [12]. What these studies showed was that spin-1 is a linear combination of vector and tensor polarizations P_V and P_T, respectively. The IUCF study used a 270 MeV, vertically polarized deuteron beam in a storage ring. The COSY study was fairly similar. The beam was fast RF-cycled through 4 polarization states to reduce systematic uncertainties: $(P_V, P_T) = (1,1), (-1,1), (0,1), (0,-2)$. These states can be flipped by bunches or extracted at an experiment point.

4. Summary

A tensor-polarized deuteron provides a spin-1 system that is a combination of quark and nuclear physics. This results in new physics opportunities, which includes new spin-structure functions. Of these, b_1 may be a good starting for an electron-ion collider. A considerable reason for this is that HERMES measures it with a complementary measurement proposed at Jefferson Lab. A polarized deuteron beam is included as part of the MEIC design. Simulation studies are underway for this. Previous studies have been carried out at the IUCF storage ring and similarly at COSY.

Journal of Physics: Conference Series **543** (2014) 012008 doi:10.1088/1742-6596/543/1/012008

References

[1] Carlson J and Schiavalla R 1988 *Rev. Mod. Phys.* **70** 743; Forrest J L *et al* 1996 *Phys. Rev. C* **54** 646
[2] Hoodbhoy P, Jaffe R L, and Manohar A 1989 *Nucl. Phys. B* **312** 571
[3] Frankfurt L L and Strikman M I 1983 *Nucl. Phys. A* **405** 557
[4] Airapetian A *et al* 2005 *Phys. Rev. Lett.* **95** 242001 Riedl C 2005 Ph.D. thesis, DESY-THESIS-2005-027
[5] Khan H and Hoodbhoy P 1991 *Phys. Rev. C* **44** 1219
[6] Umnikov A Y 1997 *Phys. Lett. B* **391** 177
[7] Edelmann J, Piller G, and Weise W 1998 *Phys. Rev. C* **57** 3392
[8] Frankfurt L, Guzey V, and Strikman M 2006 *Mod. Phys. Lett. A* **21** 23
[9] Bora K and Jaffe R L 1998 *Phys. Rev. D* **57** 6906
[10] Nikolaeva N N, Schäfer W 1997 *Phys. Lett. B* **398** 245
[11] Abeyratne S *et al* 2012 *arXiv*:1209.0757
[12] Morozov V S 2007 Ph. D. thesis, Univ. of Michigan
[13] Chen J P *et al* 2013 E12-13-011, Jefferson Lab PAC 40

Journal of Physics: Conference Series **543** (2014) 012009 doi:10.1088/1742-6596/543/1/012009

Inclusive and Exclusive Scatterings from Tensor Polarized Deuteron

M M Sargsian[1], **M I Strikman**[2]

[1] Department of Physics, Florida International University, Miami, FL 33199

[2] Department of Physics, Pennsylvania State University, University Park, PA 16802

E-mail: sargsian@fiu.edu, strikman@phys.psu.edu

Abstract. The possibility of using a tensor polarized deuteron target in electroproduction reactions creates new opportunities for studying different phenomena related to the short-range hadronic and nuclear physics. The use of the tensor polarized deuteron allows us to isolate smaller than average inter-nucleon distances for the bound two-nucleon system. In this report we consider several high Q^2 reactions which are particularly sensitive to the short-range two-nucleon configurations in the deuteron. One is the relativistic dynamics of electron-bound-nucleon scattering, which can be studied in both inclusive and exclusive reactions, and the other is the strong final state interaction in close proximity of two nucleons that can be used as a sensitive probe for color-transparency phenomena.

1. Introduction

The deuteron is the simplest nuclear system with the wave function strongly dominated by the pn component. Thus, it can be used for studies of many aspects of pn strong interactions. One such aspect is the study of the pn system at short distances where one hopes to gain access to many fundamental issues of nuclear dynamics, such as a relativistic description of nuclear structure, the dynamics of the NN repulsive core, the role of the non-nucleonic degrees of freedom, and the hadron-quark transition at very short distances.

However one problem in realizing such a program of studies is that the deuteron is barley bound with the charge-rms radius of about 2.1 fm. In the momentum space, this fact is reflected in the very steep momentum distribution of the unpolarized deuteron wave function with the strength concentrated predominantly at the small relative momenta in the pn system (see the curve labelled as ρ^{unp} in Fig.1).

However the fact that the deuteron has a D-wave which vanishes at small momenta indicates that isolating it in any given reaction with the deuteron will effectively suppress the small-momentum/long-range contributions.

This can be seen from the polarized density matrix of the deuteron [1]:

$$\rho_d^{\vec{a}}(k_1,k_2)) = u(k_1)u(k_2) + \left[1 - \frac{3|k_2\cdot a|^2}{k_2^2}\right]\frac{u(k_1)w(k_2)}{\sqrt{2}} + \left[1 - \frac{3|k_1\cdot a|^2}{k_1^2}\right]\frac{u(k_2)w(k_1)}{\sqrt{2}}$$
$$+\left(\frac{9}{2}\frac{(k_1\cdot a)(k_2\cdot a)^*(k_1\cdot k_2)}{k_1^2k_2^2} - \frac{3}{2}\frac{|k_1\cdot a|^2}{k_1^2} - \frac{3}{2}\frac{|k_2\cdot a|^2}{k_2^2} + \frac{1}{2}\right)w(k_1)w(k_2), \tag{1}$$

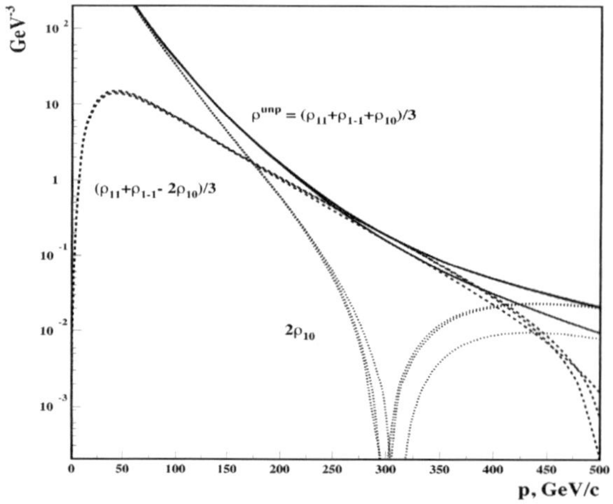

Figure 1. Momentum dependences of different combinations of the polarized density matrices. Solid, dashed and dotted curves correspond to unpolarized, tensor polarized, and transverse polarized distributions, respectively.

where $u(k)$ and $w(k)$ represent the S and D partial waves respectively. The polarization vector $\vec{a}$ is defined through the deuteron spin wave functions

$$\psi^{10} = i \cdot a_z, \; \psi^{11} = -\frac{i}{\sqrt{2}}(a_x + ia_y), \; \psi^{1-1} = \frac{i}{\sqrt{2}}(a_x - ia_y), \tag{2}$$

where $\psi^{1\mu}$ is the projection of the deuteron's spin in the μ direction. The unpolarized density matrix of the deuteron is defined as $\rho_d^{unp}(k_1,k_2) = \frac{1}{3}\sum_a \rho_d^a(k_1,k_2)$.

Since $\lim_{k\to 0} w(k) = 0$, it follows from Eq.(1) that any polarization combination of $\rho_d^{\vec{a}}$, in which the u^2 term is canceled has an enhanced sensitivity to the larger internal momenta (smaller distances) of the deuteron as compared to the unpolarized case. It follows from Eq.(1) that the $u(k_1)u(k_2)$ term does not depend on the polarization vector $\vec{a}$, thus one can cancel this term summing any two polarization components of the density matrix and subtracting the doubled value of the third polarization component. One example is:

$$\rho^{20}(k_1,k_2) \equiv \frac{1}{3}(\rho^{11} + \rho^{1-1} - 2\rho^{10}) = \left(\frac{3k_{1z}^2}{k_1^2} - 1\right)\frac{u(k_2)w(k_1)}{\sqrt{2}} + \left(\frac{3k_{2z}^2}{k_2^2} - 1\right)\frac{u(k_1)w(k_2)}{\sqrt{2}}$$
$$+ \left(\frac{3}{2}\left[\frac{(k_1k_2)(k_1k_2 - 3k_{1z}k_{2z})}{k_1^2k_2^2} + \frac{k_{1z}^2}{k_1^2} + \frac{k_{2z}^2}{k_2^2}\right] - \right) w(k_1)w(k_2). \tag{3}$$

Fig. 1 presents the examples of the density matrices for unpolarized, $((\rho_{11} + \rho_{1-1} + \rho_{10})/3$, transverse, ρ_{10}, and tensor polarized, ρ_{20}, deuteron targets as they enter in the impulse approximation term of the $\vec{d}(e,e'p)n$ cross section (in this case $k_1 = k_2 = p$). As it can be seen from Eq.(3), the tensor polarized density matrix depends only on the terms proportional to $u(p)w(p)$ and $w(p)^2$.

This suggests ways for studying several issues of nuclear physics related to the short-range interactions using tensor polarized deuteron targets.

1.1. The D-wave Component of the Deuteron

The presence of the D-wave component in the deuteron follows from the existence of the finite quadruple momentum of the deuteron. It is directly related to the tensor part of the NN interaction. The modern NN interaction potentials which fit the existing NN phase-shifts with $\xi^2 \leq 1$ predict the overall probability of the D wave in the deuteron to be in the range of 4.87% (cdBonn [2]) to 5.76% (AV18 [3]). Although the difference seems rather small, these parameterizations predict substantially different high momentum strengths above 400 MeV/c, which is predominantly related to the D component. The need to understand the D-wave momentum distribution became more pressing recently, due to the observation of the strong dominance of pn short range correlations (SRCs) in nuclei as compared to the pp and nn SRCs [4, 5], which is related to the tensor component of NN interaction in the SRC [6, 7]. The dominance of the tensor interaction in the high momentum component of the nuclear wave function may play important role in the dynamics of the asymmetric nuclei [8, 9] with nontrivial implications for superdense nuclear matter and neutron stars [10, 11].

Very recently another interesting aspect of the role of the D-wave was revealed in the observation of k^{-4} scaling of the momentum distributions in the deuteron and nuclei [12] in the momentum range of 300-600 MeV/c, which shows intriguing similarities with the contact observed in the two-component symmetric atomic systems of fermions.

As it follows from Eq.(3), the processes involving a tensor-polarized deuteron target are directly related to the structure of the D-partial wave component in the deuteron.

1.2. Relativistic Dynamics of the NN Bound System

One of the important issues in studying nuclear structure at short distances is the need for a relativistic description of the bound system. This is an important issue also in understanding the QCD medium effects with recent studies indicating that parton distribution modifications in nuclei are proportional to the high momentum component of nuclear wave function [13].

The deuteron is the simplest bound system and naturally any self-consistent attempt to understand the relativistic effects in the bound nuclear systems should start with the deuteron. The issue of the relativistic description of the deuteron has long history with extensive research started already in late 1970's (see e.g. [14, 15, 16, 17]).

The experimental studies of the relativistic effects in the unpolarized deuteron up to now included the large Q^2 elastic ed scatterings [18]. However, due to complexities in the reaction mechanism the relativistic effects were difficult to isolate.

The processes involving a tensor polarized deuteron target are expected to exhibit enhanced sensitivity to the relativistic effects due to higher average momentum of the bound nucleon entering in the polarized nuclear density matrix as compared to the unpolarized case. Thus one can discuss possible high momentum transfer reactions off tensor polarized deuterons and study their sensitivity to the relativistic effects of the bound nucleon motion.

1.3. Final Sate Hadronic Interactions

The fact that the tensor polarized deuteron is characterized by a larger average internal momenta indicates that the two nucleons are in close proximity and scattering from such target can yield large final state hadronic interactions in exclusive reactions. The latter can be used in studies of the hadronic properties of produced particles such as nucleons [1, 19, 20] or vector mesons [21, 22, 23].

2. Electroproduction Reactions Involving Tensor Polarized Deuterons

We will discuss several electroproduction reactions in which utilizing the unique features of the tensor polarized deuteron would allow us to study the above discussed properties of high energy processes.

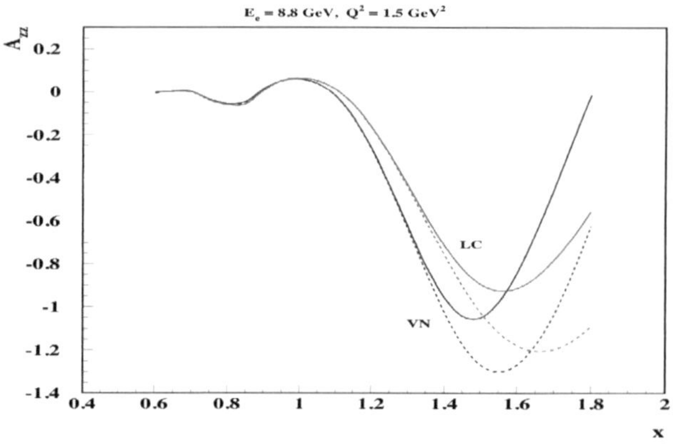

Figure 2. The x_{Bj} dependence of A_{zz} asymmetry. The "VN" and "LC" labels correspond to the VNA and and LC calculations described in the text. Solid line AV18 [3] and dashed line cdBonn [2] NN potentials are used in the calculations.

2.1. High Q^2 Near-Threshold Inclusive Scattering

High Q^2 (> 1 GeV^2) inclusive scattering at $x_{Bj} > 1$ is known to probe large longitudinal momenta of bound nucleon interacting with the virtual photon. One of the important indications that these reactions indeed probe the high momentum component of the nuclear wave function is the observation of the scaling in the ratios of the $A(e, e')X$ cross sections to that of deuteron [24, 25] or light nuclei [26, 27].

Using the tensor polarized deuteron in such reactions allows us to prepare the nucleus in the most compact state in which, due to the absence of the pure (S-wave)^2 contribution (Fig. 1), the system in average is sensitive to a higher momentum of the nucleon in the deuteron for given x and Q^2. At large $Q^2 > 1$ GeV^2 kinematics, the probed longitudinal momenta of the bound nucleon is $p_z \approx m_N(1-x)$, or the light cone momentum fraction $\alpha \geq x$. Because of these kinematic conditions and the absence of the (S-wave)^2 contribution, one expects a measurable relativistic effects already at $x \sim 1.2$. Such an early onset of the relativistic effects indicates that one should be able to separate them from the uncertainty related to the choice of the NN potential.

The sensitivity to the relativistic effects is estimated using the theoretical calculations based on two very different approaches. The first approach describes the bound nucleon in the deuteron rest frame treating the interacting nucleon as being virtual (virtual nucleon approximation (VNA)) by taking the residue over the positive energy pole of the spectator nucleon. In this case the deuteron wave function satisfies the covariant equation of a two-nucleon bound system with the spectator being on mass-shell (see e.g. [28, 29]).

Another approach is based on the observation that high energy processes evolve along the light-cone. Therefore, it is natural to describe the reaction within the light-cone non-covariant framework [16, 17]. Negative energy states do not enter in this case, though one has to take into account so called instantaneous interactions. In the approximation when non-nucleonic degrees of freedom in the deuteron wave function can be neglected, one can relate the light-cone wave functions to those calculated in the lab frame by introducing the LC pn relative three momentum $k = \sqrt{\frac{m^2+p_t^2}{\alpha(2-\alpha)} - m^2}$.

In Fig. 2 the predictions for VNA [28] and LC [24] approximations are given for the tensor asymmetry, $A_{zz} = \frac{T^{20}}{\sigma^{unp}}$, at the highest Q^2 kinematics for the recently proposed experiment at Jefferson Lab [30]. As the figure shows for $1.2 \leq x \leq 1.4$ the uncertainty due to the NN potentials used in the calculations is much smaller than the relativistic effects, while at high $x > 1.4$ we have substantial uncertainty due to the potentials. The calculations predict large A_{zz} asymmetry in $x > 1$ region with significant potential of discrimination between both relativistic and NN potential effects.

2.2. High Momentum Transfer Exclusive Electrodisintegration of the Deuteron

The possibility of the extension of inclusive measurements to the exclusive electrodisintegration reactions in which struck nucleon carries almost all the momentum of the virtual nucleon gives direct access to the dynamics of the bound nucleon. Recently the first high Q^2 experiments were completed for $d(e, e'p)n$ reactions [31], in which it was observed that due to the onset of the eikonal regime in the final state interaction (FSI) of the stuck proton with the spectator neutron it is possible to isolate kinematic regions with minimal and maximal FSI effects. This observation is in agreement with the theoretical calculations which are based on high energy approximations in the description of FSI effects (see e.g. [1, 19, 20, 32, 28, 33]).

The pattern of FSI in $d(e, e'p)n$ reaction is best seen by considering the ratio of the cross section of the full calculation to that of the plane wave impulse approximation: $T = \frac{\sigma^{Full}}{\sigma^{PWIA}}$.

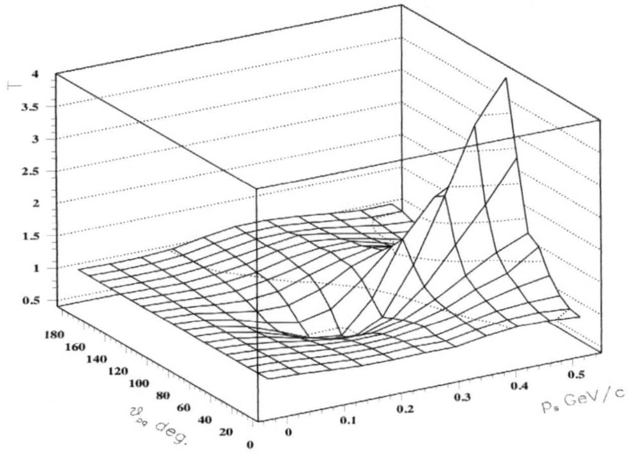

Figure 3. The dependence of the transparency T on the angle - θ_{sq} and the momentum, p_s of the recoil nucleon. The angle is defined relative to **q**.

Fig. 3 presents the expectations for T as a function of the recoil nucleon angle θ_{sq} relative to $\vec{q}$ for different values of recoil nucleon momentum. The figure demonstrates the distinctive angular dependence of the ratio T. At recoil nucleon momenta $p_s \leq 300 MeV/c$, $T \leq 1$ and has a minimum, while at $p_s > 300 MeV/c$, $T > 1$ and has a distinctive maximum. It can be seen from this picture that the FSI is small at kinematics in which recoil momenta in the reaction is parallel or anti-parallel to **q** (referred to as collinear kinematics). The FSI dominates in the kinematics where $\theta_{pq} \approx 90^0$, more precisely the maximal re-scattering corresponds to the kinematics in which $\alpha \equiv \frac{E_s - p_{sz}}{m} = 1$ (referred to as transverse kinematics). The analysis of Fig. 3 shows that one can indeed isolate the kinematic domains where PWIA term is dominant from the domain in which FSI plays a major role. The ability to identify these two kinematics is an important advantage of high Q^2 $d(e, e'p)n$ reactions. It allows us to concentrate on the different aspects of the dynamics of $d(e,e'p)n$ reaction with less background effects. Namely the collinear kinematics are best suited for studies of bound nucleon dynamics, while in transverse kinematics one can concentrate on the physics of hadronic re-interaction.

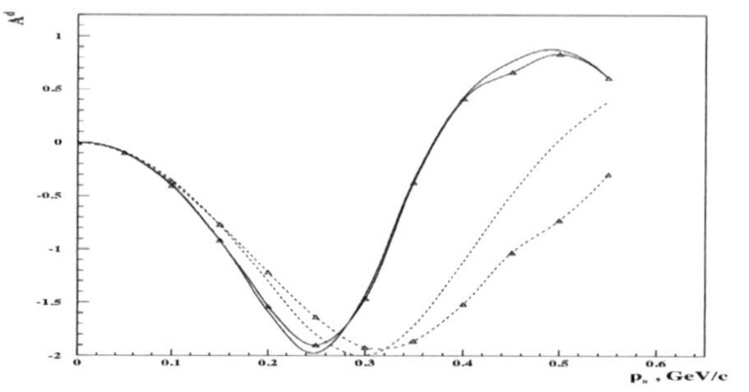

Figure 4. The p_s dependence of the $d(e,e'p)n$ tensor polarization asymmetry at $\theta_{sq} = 180^0$. Solid and dashed lines are PWIA predictions of the LC and VNA approximations. The marked curves present the same (LC, VNA) calculations including the FSI effects.

To quantify these statements, in Fig. 4 we present the calculation of A_d (which is similar to A_{zz} but for exclusive processes) in collinear kinematics ($\theta_{sq} = 180^o$) within virtual nucleon and light-cone approximations described above. These estimates, similar to the inclusive scattering case, show significant relativistic effects already at moderate momenta of $250 - 300$MeV/c for the spectator nucleon. It is interesting that the FSI effects further increase the difference between VNA and LC predictions, making it even easier to discriminate between VNA and LC approximations.

2.3. Dynamics of the Final State Interactions

As it follows from Fig. 3, studying the $d(e,e'p)n$ reactions at transverse kinematics allows us to enhance the effects due final state interaction of the struck nucleon off the spectator nucleon. One expects that FSI effects will be further enhanced for tensor-polarized deuterons since it corresponds, on average, to a smaller configuration than in the case of the unpolarized target.

If one assumes that FSI are determined by diffractive small angle pn rescatterings, one can estimate the effect of the FSI on A_d, which is presented in Fig. 5.

The comparison of Fig. 5(a) and (b) shows the very strong sensitivity of the asymmetry A_d to the final state interaction at the transverse kinematics: the FSI significantly diminishes the asymmetry starting at $p_n = 300$ MeV/c and $\theta_{nq} \sim 90^0$ (or $\alpha_n \approx 1$). This indicates that the quantity A_d can be used as a very sensitive observable for studying the FSI dynamics; for example, the onset of the Color Transparency (CT) phenomena (for details of CT phenomena see Ref. [34]) at large Q^2 (≥ 4 GeV^2).

For numerical estimates, we consider the Q^2 dependence of the asymmetry A_d for fixed and transverse momenta of the spectator neutron. This dependence for $p_t = 300$ MeV/c, is presented in Fig. 6. One can see from this figure that CT effects can change A_d by as much as factor of two for $Q^2 \sim 10$ GeV^2. It is worth noting that the same models predict only 10-15% effect for $(e,e'p)$ reactions on unpolarized nuclear targets.

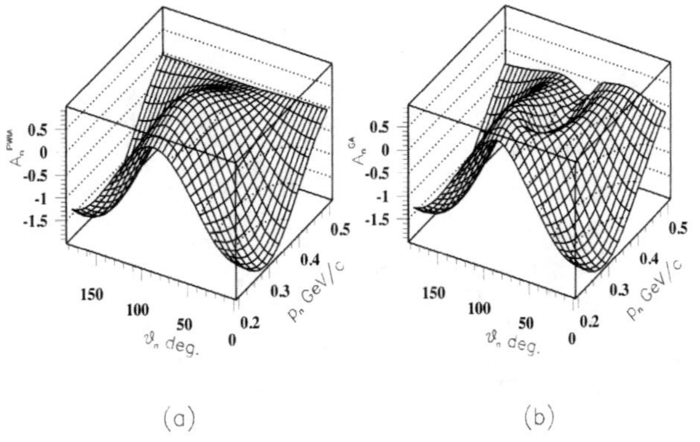

Figure 5. The dependence of tensor asymmetry A_d on the angle and momentum of the spectator neutron in $d(e, e'p)n$ reaction. (a) PWIA approximation (b) Full calculation that includes FSI.

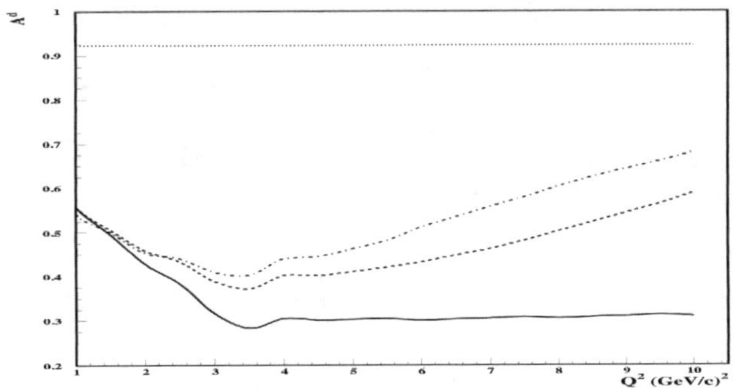

Figure 6. The Q^2 dependence of A_d for $\alpha_n = 1$. Solid line - Full calculation without CT effects, dashed prediction of Quantum Diffusion Model [35], dashed-dotted - prediction of three state resonance model model [36], dotted -PWIA.

3. Conclusion and Outlook

The use of the tensor polarized deuteron targets in high energy electro-production reactions would provide a unique possibility for studying the NN strong interaction at short space-time separations. We demonstrate that such reactions allow a direct test of the relativistic effects in the short range NN interaction. They also allow us to discriminate between different NN potentials.

Another advantage of the tensor polarized deuteron as a compact two-nucleon system is the possibility to significantly enhance the final state hadronic interaction in the exclusive $d(e, e', p)n$ reactions. This provides a very sensitive tool for studying color transparency phenomena, which will be manifested in the suppression of the FSI effects at large Q^2 kinematics.

All these studies are a part of the broad program dedicated to the investigation of the structure of deeply bound nucleons as well as the physics of color transparency [37]. This program could benefit tremendously from the advances of building polarized deuteron targets that can be operated with high current electron beams.

Acknowledgements
This work is supported by United States DOE grants under contract DE-FG02-01ER-41172 and DE-FG02-93ER40771.

References

[1] Frankfurt L L, Greenberg W R, Miller G A, Sargsian M M, Strikman M I 1995 *Z. Phys.* **A352** 97; Frankfurt L L, Greenberg W R, Miller G A, Sargsian M M, Strikman M I 1996 *Phys. Lett.* **B369** 201
[2] Machleidt R 2001 *Phys. Rev.* **C63** 024001
[3] Wiringa R B, Stoks V G J, and R. Schiavilla 1995 *Phys. Rev.* **C51** 38
[4] Piasetzky E, Sargsian M, Frankfurt L, Strikman M and Watson J W 2006 *Phys. Rev. Lett.* **97** 162504
[5] Subedi R, Shneor R, Monaghan P *et al* 2008 *Science* **320** 1476
[6] Sargsian M M, Abrahamyan T V, Strikman M I, and Frankfurt L L 2005 *Phys. Rev.* **C71** 044615
[7] Schiavilla R, Wiringa R B, Pieper S C, and Carlson J 2007 *Phys. Rev. Lett.* **98** 132501
[8] Sargsian M M 2014 *Phys. Rev.* **C89** 034305
[9] McGauley M and Sargsian M M 2011 *arXiv*:1102.3973 [nucl-th]
[10] Frankfurt L, Sargsian M and Strikman M 2008 *Int. J. Mod. Phys. A* **23** 2991
[11] Arrington J, Higinbotham D W, Rosner G, and Sargsian M 2012 *Prog. Part. Nucl. Phys.* **67** 898
[12] Hen O, Weinstein L B, Piasetzky E, Miller G A, and Sargsian M M 2014 *arXiv*:1407.8175 [nucl-ex]
[13] Weinstein L B, Piasetzky E, Higinbotham D W, Gomez J, Hen O, and Shneor R 2011 *Phys. Rev. Lett.* **106** 052301
[14] Frankfurt L L and Strikman M I 1979 *Nucl. Phys. B* **148** 107
[15] Buck W W and Gross F 1979 *Phys. Rev.* **D20** 2361
[16] Frankfurt L L and Strikman M I 1981 *Phys. Rep.* **76** 217
[17] Frankfurt L L and Strikman M I 1988 *Phys. Rep.* **160** 235
[18] Alexa L C *et al* [Jefferson Lab Hall A Collaboration] 1999 *Phys. Rev. Lett.* **82** 1374
[19] Frankfurt L L, Sargsian M M, Strikman M I 1997 *Phys. Rev.* **C56** 1124
[20] Sargsian M M 2001 *Int. J. Mod. Phys. E* **10** 405
[21] Frankfurt L, Koepf W, Mutzbauer J, Piller G, Sargsian M and Strikman M 1997 *Nucl. Phys. A* **622** 511
[22] Frankfurt L, Piller G, Sargsian M and Strikman M 1998 *Eur. Phys. J. A* **2** 301
[23] Freese A J and Sargsian M M 2013 *Phys. Rev.* **C88** 044604
[24] Frankfurt L L, Strikman M I, Day D B and Sargsian M 1993 *Phys. Rev.* **C48** 2451
[25] Fomin N, Arrington J, Asaturyan R *et al* 2012 *Phys. Rev. Lett.* **108** 092502
[26] Egiyan K S *et al* [CLAS Collaboration] 2003 *Phys. Rev.* **C68** 014313
[27] Egiyan K S *et al* [CLAS Collaboration] 2006 *Phys. Rev. Lett.* **96** 082501
[28] Sargsian M M 2012 *Phys. Rev.* **C82** 014612
[29] Gross F and Stadler A 2010 *Phys. Rev.* **C82** 034004
[30] Long E, Slifer K, Solvignon P, Day D, Keller D, Higinbotham D *et al* 2014 *Tensor Asymmetry A_{zz} in the $x>1$ Region* JLab LOI-12-14-002
[31] Boeglin W U *et al* [Hall A Collaboration] 2011 *Phys. Rev. Lett.* **107** 262501
[32] Jeschonnek S and Van Orden J W 2008 *Phys. Rev.* **C78** 014007
[33] Ford W P, Jeschonnek S, and Van Orden J W 2013 *Phys. Rev.* **C87** 054006
[34] Frankfurt L L, Miller G A, and Strikman M I 1994 *Ann. Rev. Nucl. Part. Phys.* **44** 501
[35] Farrar G R, Frankfurt L L, Strikman M I, and Liu H 1988 *Phys. Rev. Lett.* **61** 686
[36] Frankfurt L L, Greenberg W R, Miller G A, and Strikman M I 1992 *Phys. Rev.* **C46** 2547
[37] Sargsian M M, Arrington J, Bertozzi W *et al* 2003 *J. Phys. G* **29** R1

Journal of Physics: Conference Series **543** (2014) 012010 doi:10.1088/1742-6596/543/1/012010

Potential for a Tensor Asymmetry A_{zz} Measurement in the $x > 1$ Region at Jefferson Lab

E Long

Post Doctoral Research Associate, University of New Hampshire, Durham, NH, 03824, USA

E-mail: ellie@jlab.org

Abstract. The tensor asymmetry A_{zz} in the quasi-elastic region through the tensor polarized $D(e, e')X$ channel is sensitive to the nucleon-nucleon potential. Previous measurements of A_{zz} have been used to extract b_1 in the DIS region and T_{20} in the elastic region. In the quasi-elastic region, A_{zz} can be used to compare light cone calculations with variation nucleon-nucleon calculations, and is an important quantity to determine for understanding tensor effects, such as the dominance of pn correlations in nuclei. In the quasi-elastic region, A_{zz} was first calculated in 1988 by Frankfurt and Strikman using the Hamada-Johnstone and Reid soft-core wave functions [1]. Recent calculations by M. Sargsian revisit A_{zz} in the $x > 1$ range using virtual-nucleon and light-cone methods, which differ by up to a factor of two [2]. Discussed in these proceedings, a study has been completed that determines the feasibility of measuring A_{zz} in the quasi-elastic $x > 1$ region at Jefferson Lab's Hall C.

1. Background

It was suggested for some time [3] that to resolve the microscopic structure of nuclei one needs to study scattering at sufficiently large momentum transfer and large relative momenta of the produced nucleons. This logic was confirmed [4] by a series of experiments at SLAC [5] and JLab [6, 7] that directly observed short-range correlations (SRC) in a series of nuclei, and established a similar effect of SRC in the deuteron and in heavier nuclei with pn correlations giving the dominant contribution. Hence, the deuteron serves as a "hydrogen atom" for the studies of the microscopic short-range structure of the nuclei since it is the simplest nuclei that follows SRC scaling. To achieve further progress, it is necessary to improve our knowledge of the deuteron wave function and nucleon-nucleon interactions at high momenta.

Due to their small size and simple structure, tensor polarized deuterons are ideal for studying nucleon-nucleon interactions. Tensor polarization enhances the D-state contribution and is sensitive to short-range QCD effects [8]. Understanding the nucleon-nucleon potential of the deuteron is essential for understanding short-range correlations, as they are largely dependent on the tensor force [4].

By taking a ratio of cross sections of electron scattering from tensor-polarized and unpolarized deuterons, the S and D-wave states can be disentangled, leading to a fuller understanding of the repulsive nucleon core. A measurement of A_{zz} is sensitive to the $\frac{D^2-SD}{S^2+D^2}$ ratio and its evolution with increasing minimal momentum of the struck nucleon. Originally calculated by L. Frankfurt and M. Strikman [1], this has recently been revisited by M. Sargsian who calculated A_{zz} at $Q^2 = 1.5$ (GeV$/c)^2$ and $0.6 < x < 1.8$ using both a light cone and virtual nucleon approach [2].

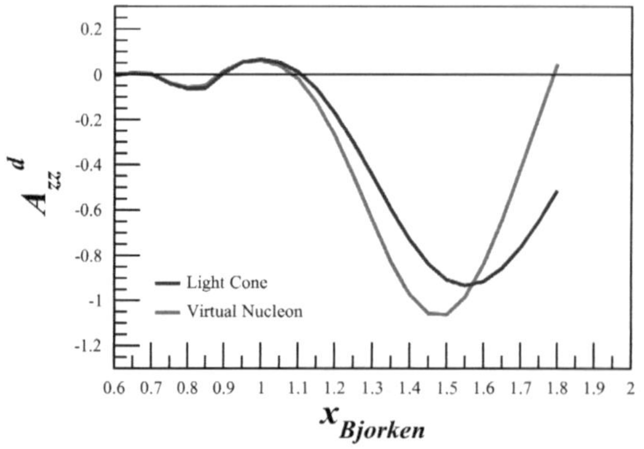

Figure 1. The A_{zz} observable calculated at $Q^2 = 1.5$ (GeV/c)^2 using the light-cone and virtual nucleon models. Calculations provided by M. Sargsian.

The calculations vary by up to a factor of 2, which, combined with the 12 GeV Jefferson Lab upgrade and the new Super High Momentum Spectrometer (SHMS) to be installed in Hall C, make A_{zz} an ideal candidate to revisit experimentally.

Additionally, measuring A_{zz} in the quasi-elastic region will fill a gap in measurements performed on tensor polarized deuterons. In the elastic region, A_{zz} is directly proportional to T_{20} ($A_{zz} \propto T_{20}$). In the deep inelastic region, A_{zz} will soon be measured to extract the tensor structure function b_1 by the relation $A_{zz} \propto \frac{b_1}{F_1^D}$. Not only would measuring A_{zz} in the quasi-elastic region provide information necessary for understanding the high-momentum properties of the deuteron and contributions from the tensor force, but it would be the first experiment to bridge a gap in measurements of electron scattering from tensor-polarized deuterons.

2. Feasibility Study

We discuss here the feasibility of a measurement of the tensor asymmetry A_{zz} from inclusive electron scattering from polarized deuterons in the quasi-elastic region of $0.80 < x < 1.75$, $0.3\ (\text{GeV}/c)^2 < Q^2 < 1.5\ (\text{GeV}/c)^2$, and $0.59 < W < 1.09$ GeV at Jefferson Lab using a DNP solid polarized ND_3 target and the Hall C HMS and SHMS spectrometers at forward angles.

2.1. Definitions

The measured double differential cross section for a spin-1 target is characterized by a vector polarization, P_z, and tensor polarization, P_{zz}, expressed as,

$$\frac{d^2\sigma_p}{d\Omega dE'} = \frac{d^2\sigma_u}{d\Omega dE'}\left(1 - P_z P_B A_1 + \frac{1}{2}P_{zz}A_{zz}\right), \tag{1}$$

where, σ_p (σ_u) is the polarized (unpolarized) cross section, P_B is the incident electron beam polarization, and A_1 (A_{zz}) is the vector (tensor) asymmetry of the virtual-photon deuteron cross section. This allows us to write the polarized tensor asymmetry with positive tensor polarization using an unpolarized electron beam as

$$A_{zz} = \frac{2}{P_{zz}}\left(\frac{\sigma_p - \sigma_u}{\sigma_u}\right). \tag{2}$$

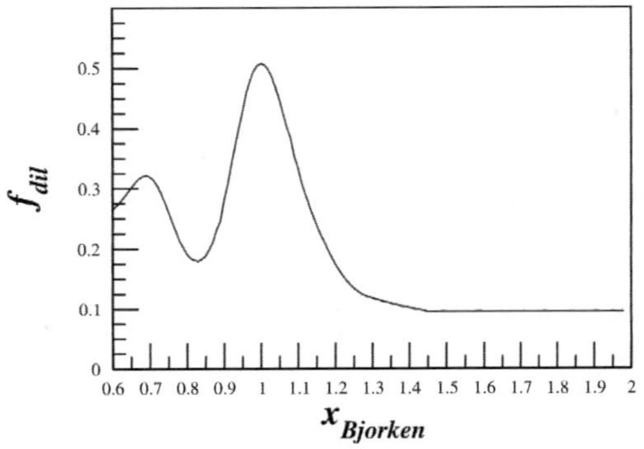

Figure 2. The estimated dilution factor, in this case at $Q^2 = 1.5\ (\mathrm{GeV}/c)^2$, drops off at high x until it reaches the SRC plateau region.

The tensor polarization is given by

$$P_{zz} = \frac{n_+ - 2n_0 + n_-}{n_+ + n_- + n_0}, \tag{3}$$

where n_m represents the population in the $m_j = +1, -1$, or 0 state.

Eq. 2 reveals that the asymmetry A_{zz} compares two different cross sections measured under different polarization conditions of the target: positively tensor polarized and unpolarized. To obtain the relative cross section measurement in the same configuration, the same target cup and material will be used at alternating polarization states (polarized vs. unpolarized), and the magnetic field providing the quantization axis will be oriented along the beamline at all times. This field will always be held at the same value, regardless of the target material polarization state. This process ensures that the acceptance remains consistent within the stability (10^{-4}) of the super conducting magnet.

Since many of the factors involved in the cross sections cancel in the ratio, Eq. 2 can be expressed in terms of the charge normalized and efficiency corrected numbers of tensor polarized (N_p) and unpolarized (N_u) counts,

$$A_{zz} = \frac{2}{fP_{zz}}\left(\frac{N_p - N_u}{N_u}\right). \tag{4}$$

The dilution factor f corrects for the presence of unpolarized nuclei in the target and is defined by

$$f = \frac{N_D\sigma_D}{N_N\sigma_N + N_D\sigma_D + \sum_A N_A\sigma_A}, \tag{5}$$

where N_D is the number of deuterium nuclei in the target and σ_D is the corresponding inclusive double differential scattering cross section, N_N is the number of nitrogen nuclei with cross section σ_N, and N_A is the number of other scattering nuclei of mass number A with cross section σ_A. As has been noted in previous work [1], the dilution factor at high x drops off considerably until the SRC plateau region, as shown in Fig. 2. Although this effect makes A_{zz} in the large x region difficult to measure at large Q^2, the Jefferson Lab 12 GeV upgrade and the new SHMS spectrometer provide the experimental infrastructure necessary to overcome it.

The dilution factor can be written in terms of the relative volume ratio of ND_3 to LHe in the target cell, otherwise known as the packing fraction p_f. In the case of a cylindrical target cell

Journal of Physics: Conference Series **543** (2014) 012010 doi:10.1088/1742-6596/543/1/012010

Source	Systematic
P_{zz} Polarimetry	12%
Dilution Factor	6.0%
Packing Fraction	3.0%
Trigger/Tracking Efficiency	1.0%
Acceptance	0.5%
Charge Determination	1.0%
Detector Resolution and Efficiency	1.0%
Total	14%

Table 1. Estimates of the scale dependent contributions to the systematic error of A_{zz}.

oriented along the magnetic field, the packing fraction is exactly equivalent to the percentage of the cell length filled with ND_3.

If the time is evenly split between scattering off of polarized and unpolarized ND_3, the time necessary to achieve the desired precision δA is

$$T = \frac{N_p}{R_p} + \frac{N_u}{R_u} = \frac{8}{f^2 P_{zz}^2}\left(\frac{R_p(R_u + R_p)}{R_u^3}\right)\frac{1}{\delta A_{zz}^2}, \tag{6}$$

where $R_{p(u)}$ is the polarized (unpolarized) rate and $N_{p(u)}$ is the total estimated number of polarized (unpolarized) counts to achieve the uncertainty δA_{zz}.

2.2. *Statistical Uncertainty Estimate*

To investigate the statistical uncertainty, we start with the equation for A_{zz} using measured counts for polarized data (N_p) and unpolarized data (N_u),

$$A_{zz} = \frac{2}{fP_{zz}}\left(\frac{N_p}{N_u} - 1\right). \tag{7}$$

The statistical error with respect to counts is then

$$\delta A_{zz} = \frac{2}{fP_{zz}}\sqrt{\left(\frac{\delta N_p}{N_u}\right)^2 + \left(\frac{N_p \delta N_u}{N_u^2}\right)^2}. \tag{8}$$

For $\delta N_{p(u)} = \sqrt{N_{p(u)}}$, the uncertainty becomes

$$\delta A_{zz} = \frac{2}{fP_{zz}}\sqrt{\frac{N_p(N_u + N_p)}{N_u^3}}, \tag{9}$$

which isn't simplified further due to the expected large asymmetry.

The number of counts was calculated using a combination of P. Bosted's [9] and M. Sargsian's [10] cross-section calculation code. The Bosted code was used for the lowest Q^2 setting, where effects of SRC scaling are expected to be negligible, and for $x < 1.1$ to accurately determine the quasi-elastic peak. The Sargsian code was used for the higher Q^2 settings at $x > 1.1$ due to its inclusion of SRC scaling effects that represent a more accurate dilution factor.

Spectrometer Setting		E_0 (GeV)	Q^2 (GeV^2)	E' (GeV)	$\theta_{e'}$ (°)	Rates (kHz)
SHMS	A	8.8	1.5	8.36	8.2	0.43
SHMS	B	6.6	0.7	6.50	8.2	3.19
SHMS	C	2.2	0.3	2.11	14.4	3.73
HMS	C	2.2	0.3	2.11	14.9	2.92

Table 2. Summary of the central kinematics and trigger rates expected using the Jefferson Lab Hall C spectrometers at three settings (A, B, and C).

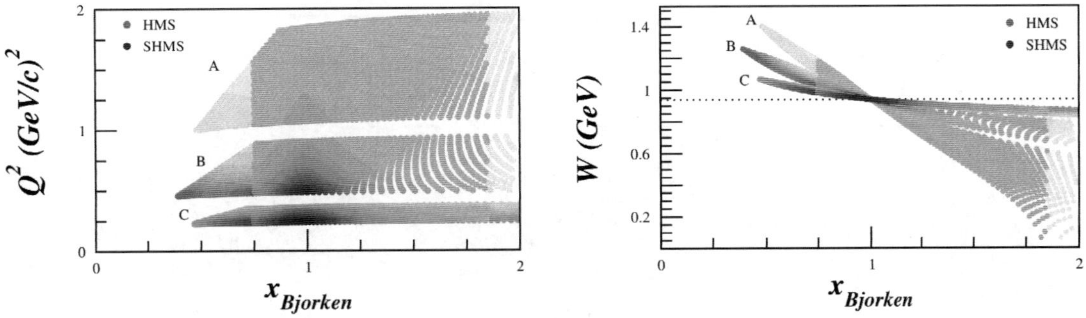

Figure 3. Kinematic coverage for central spectrometer settings at $Q^2 = 1.5$ (GeV/c)^2 (A), 0.7 (GeV/c)^2 (B), and 0.3 (GeV/c)^2 (C). The HMS is only used for setting C, and its coverage largely falls under the SHMS coverage. The grey regions are not included in our statistics estimates since they fall outside of $0.80 < x < 1.75$. Darker shading represents areas with higher statistics, and the dotted line in the W plot indicates nucleon mass.

2.3. Systematic Uncertainty Estimate

Table 1 shows a list of the scale dependent uncertainties contributing to the systematic error in A_{zz}. With careful uncertainty minimization in polarization, the relative error in vector polarization can be less than or equal to 3.9%, as was demonstrated for the proton in the recent E08-027/E08-007 experiment [11] and nearly as good for the deuteron using multiple techniques to measure the NMR signal as discussed in [12]. With the use of a positive tensor enhanced target, it has been projected to be able to achieve a relative error in P_{zz} better than 12% [12]. The uncertainty from the dilution in the polarized target is estimated to be about 6% over the range of kinematic points of interest. We consider separately the uncertainty in the packing fraction of the ammonia target to contribute at a level of less than 3%. Charge calibration and detector efficiencies are expected to be known better to 1%.

2.4. Kinematics

Utilizing the 12 GeV Jefferson Lab upgraded equipment in Hall C, the tensor asymmetry A_{zz} can be measured for $0.80 < x < 1.75$, $0.3\ (\text{GeV}/c)^2 < Q^2 < 1.5\ (\text{GeV}/c)^2$, and $0.59 < W < 1.09$ GeV. Fig. 3 demonstrates the potential kinematic coverage of such a measurement.

The experiment would use a solid tensor-polarized ND_3 target with the magnetic field held constant along the beamline. The target state would be alternated between a polarized and unpolarized state to measure A_{zz}. The tensor polarization and packing fraction used in the rates estimate are 30% and 0.65, respectively. The dilution fraction in the range of this measurement

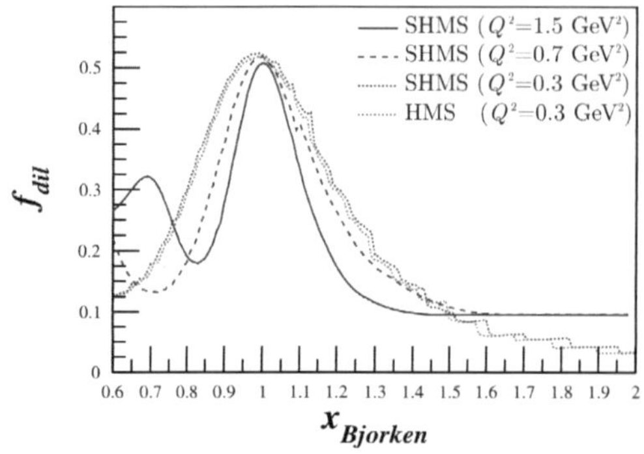

Figure 4. Projected dilution factor covering the entire x range to be measured.

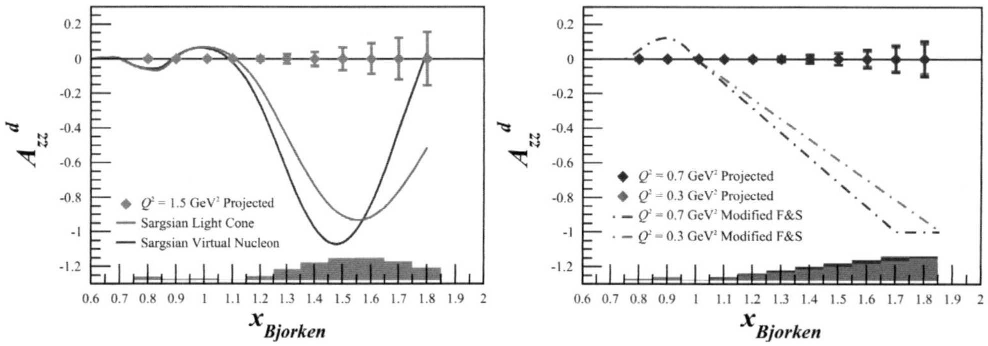

Figure 5. Projected statistical errors for the tensor asymmetry A_{zz} with 30 days of beam time. The band represents the systematic uncertainty. Also shown are calculations provided by M. Sargsian for using a light cone and virtual nucleon model for $Q^2 = 1.5$ (GeV/c)^2, and a modified Frankfurt and Strikman model [1] that estimates the peak shifts in x expected due to the SRC scaling changing with Q^2 [13] for $Q^2 = 0.3$ and 0.7 (GeV/c)^2.

is shown in Fig. 4. With an incident electron beam current of 90 nA, the expected deuteron luminosity is 1.3×10^{35} cm^{-2}s^{-1}.

The momentum bite and the acceptance were assumed to be $\Delta P = \pm 8\%$ and $\Delta\Omega = 5.6$ msr for the HMS, and $\Delta P =^{+20\%}_{-8\%}$ and $\Delta\Omega = 4.4$ msr for the SHMS. For the choice of the kinematics, special attention was taken into the angular and momentum limits of the spectrometers: for the HMS, $10.5° \leq \theta \leq 85°$ and $1 \leq P_0 \leq 7.3$ GeV/c, and for the SHMS, $5.5° \leq \theta \leq 40°$ and $2 \leq P_0 \leq 11$ GeV/c. In addition, the opening angle between the spectrometers is physically constrained to be larger than 17.5°. The projected uncertainties in A_{zz} are shown in Fig. 5.

3. Summary

We have investigated the possibility of making high precision measurements of the quasi-elastic tensor asymmetry A_{zz} and found the opportunity compelling. By covering the kinematic range from the QE peak ($x = 1$) up to elastic scattering ($x = 2$), we expect that this data will provide valuable new insights about the high momentum components of the deuteron wavefunction.

We have found that A_{zz} can be measured with high precision at $Q^2 = 1.5$, 0.7, and 0.3 (GeV/c)^2 at Jefferson Lab in Hall C using identical equipment as the upcoming b_1 experiment

E12-13-011. Such a measurement would provide valuable insights into the NN potential and fill a gap in measurements of A_{zz} between the $T_{20} \propto A_{zz}$ elastic measurements and the $b_1 \propto \frac{A_{zz}}{F_1^d}$ deep-inelastic measurements.

4. References

[1] Frankfurt L and Strikman M 1988 *Phys.Rept.* **160** 235–427
[2] Sargsian M private communication, to be published
[3] Frankfurt L L and Strikman M I 1981 *Phys. Rept.* **76** 215–347
[4] Arrington J, Higinbotham D, Rosner G and Sargsian M 2012 *Prog.Part.Nucl.Phys.* **67** 898–938 (*Preprint* 1104.1196)
[5] Frankfurt L, Strikman M, Day D and Sargsian M 1993 *Phys.Rev.* **C48** 2451–2461
[6] Arrington J, Armstrong C, Averett T, Baker O K, de Bever L *et al.* 1999 *Phys.Rev.Lett.* **82** 2056–2059 (*Preprint* nucl-ex/9811008)
[7] Fomin N, Arrington J, Asaturyan R, Benmokhtar F, Boeglin W *et al.* 2012 *Phys.Rev.Lett.* **108** 092502 (*Preprint* 1107.3583)
[8] Forest J L *et al.* 1996 *Phys. Rev.* **C54** 646–667 (*Preprint* nucl-th/9603035)
[9] Bosted P and Mamyan V 2012 *e-print* **arXiv:1203.2262** (*Preprint* 1203.2262)
[10] M. Sargsian, private communication
[11] Keller D 2013 *Nucl. Inst. and Meth.* **A728** 133–144
[12] Keller D 2013 *XVth International Workshop on Polarized Sources, Targets, and Polarimetry* **PoS(PSTP2013)010**
[13] Frankfurt L, Sargsian M and Strikman M 2008 *Int.J.Mod.Phys.* **A23** 2991–3055 (*Preprint* 0806.4412)

Journal of Physics: Conference Series **543** (2014) 012011 doi:10.1088/1742-6596/543/1/012011

Electro-Disintegration of Tensor Polarized Deuterium

W U Boeglin

Dept. of Physics, Florida International University, Miami, FL 33199

E-mail: boeglinw@fiu.edu

Abstract. A tensor polarized target in Hall A at Jefferson Lab would offer the possibility to measure the D(e,e′p)n cross section for the $M_s = 0$ and the $M_s = \pm 1$ states separately (the quantization axis is along the momentum transfer). These data would serve as a new, stringent test of our current understanding of the deuteron structure for missing momenta up to 450 MeV/c, a region where the deuteron wave function is dominated by the D-state. No data exist to date for missing momenta above 150 MeV/c. The technique to separate these cross sections, possible kinematic settings, and a rough estimate of the achievable precision is presented.

1. Introduction

The deuteron plays a fundamental role in nuclear physics similar to the one of the hydrogen atom in atomic physics. Within modern models using realistic nucleon-nucleon (NN) interactions, one can calculate the deuterons bound and continuum states with great precision. This in turn makes it possible to predict a wide range of observables that are sensitive to details of the different aspects of the deuteron structure and indirectly to details of the NN interaction. A better knowledge of the short-range structure of the deuteron is of fundamental importance for the advancement of our understanding of the poorly known short-range structure of the nucleon-nucleon interaction as well as the theory of high-density nuclear matter where short-range structures play a crucial role. This has been illustrated by recent double and triple coincidence experiments on carbon that show that high momentum protons are predominantly paired with high momentum neutrons in np short range correlations (SRC) [1, 2, 3]. High momentum in this case means that the nucleon momentum is above the typical nuclear Fermi momentum of $k_F \approx 250$ MeV/c. This np dominance of the SRC is thought to be due to the dominance of the tensor part of the nucleon-nucleon interaction [4, 5, 6].

Probing the short-range structure of the deuteron means that one is investigating configurations where the two bound nucleons are essentially overlapping. One of the fundamental questions related to these configurations is the extent to which they can be described purely in terms of two nucleons with high initial relative momenta and whether current state-of-the-art models agree with experiment. To shed light on this problem, experimental data are needed on reactions that are sensitive to high-momentum configurations and which can also be accurately modeled. Traditionally, three classes of reactions are used for deuteron studies: elastic electron scattering, inclusive quasi-free electro-disintegration and exclusive quasi-free electro-disintegration. While elastic electron scattering has been carried out to very high momentum transfers, non-nucleonic contributions dominate the form factor and are therefore masking the underlying short range deuteron structure. The inclusive quasi-free electron scattering cross section is related to integrals of the deuteron momentum distribution within kinematic limits

Journal of Physics: Conference Series **543** (2014) 012011 doi:10.1088/1742-6596/543/1/012011

determined by energy and momentum conservation making the extraction of the momentum distribution challenging. The most direct access to the deuteron momentum distribution is via the exclusive electro-disintegration. If one assumes that the ejected proton after it absorbed the virtual photon does not interact with the recoiling neutron it can be described by a plane wave and the cross section within the so-called plane wave impulse approximation (PWIA) can be written as follows:

$$\frac{\mathrm{d}^5\sigma}{\mathrm{d}\omega \mathrm{d}\Omega_e \mathrm{d}\Omega_p} = \kappa \cdot \sigma_{ep} \cdot \rho(p_m) \tag{1}$$

Here $\rho(p_m)$ is the momentum distribution with p_m as the initial nucleon momentum, σ_{ep} is the (off-shell) electron nucleon cross section and κ is a kinematic factor. In reality, PWIA does not provide a complete description of the D(e,e′p)n reaction and final state interactions (FSI) accounting for the interaction of the outgoing nucleon with the residual system and non-nucleonic degrees of freedoms such as meson exchange currents (MEC) and isobar configurations (IC) need to be included. Indeed, D(e,e′p)n experiments at lower momentum transfers have shown that for missing momenta $p_m > 0.25 - 0.3$ GeV/c FSI quickly dominate the cross section and therefore mask details of the underlying deuteron structure [7]. Also contributions of MEC and IC are very sensitive to the reaction kinematics and in general increase with decreasing momentum transfer.

As the $S = 1$ and $T = 0$ deuteron ground state has a quadrupole moment its ground state wave function must contain a D-state contribution. This admixture is entirely due to the tensor force of the NN interaction. Realistic calculations of the wave function show that the D-state dominates the momentum distribution for missing momenta between 0.3 and 0.6 GeV/c. Hence, experimental data that are sensitive to details of the momentum distribution in this nucleon momentum range will provide important constraints on this aspect of the interaction and on the structure of short range correlations in nuclei.

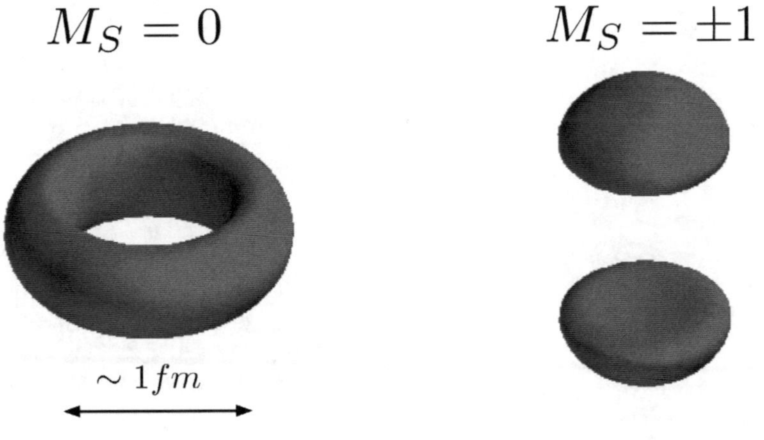

Figure 1. The surfaces with $\rho = 0.3fm^{-3}$ for $M_s = 0$ (left) and $M_s = \pm 1$ (right) from reference [8].

The short range structure of the ground state wave function is especially influenced by the repulsive core and the tensor part of the nucleon-nucleon interaction. A detailed theoretical study [8] of the two nucleon density distribution in deuterium and heavier nuclei has shown that the two nucleon distributions in the $T = 0$ isospin and $S = 1$ spin state have a strong dependence on the spin projection M_s. If the two nucleons are in a relative $M_s = 0$ state, the surface of constant density has the shape of a toroid while if the two nucleons are found in the $M_s = \pm 1$ state the surface of constant density has a dumbbell shape (Fig. 1). The two nucleon density is at a maximum in the $M_S = 0$ state for a torus with a diameter of approximately 1 fm and has a value of $\approx 0.34\ fm^{-3}$ which corresponds to twice nuclear matter density. The shape of the torus is produced by the combined action of the tensor force and the repulsive core which is responsible for the hole.

Determining the D(e,e′p)n cross section for tensor polarized deuterium would provide very direct data to test quantitatively our understanding of this important aspect of the NN interaction and lead to new information regarding the limit of a conventional description of nuclei in terms of pure nucleonic degrees of freedom. More specifically, this new information

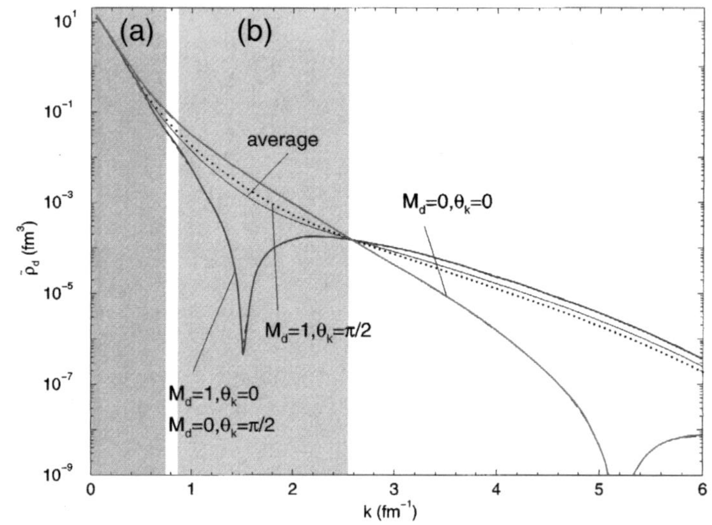

Figure 2. The calculated deuteron momentum distribution for different values of M_S and θ_k from reference [8]. The area (a) indicates the missing momentum range covered by the NIKEF experiment [9] and area (b) represents the kinematic range that could be explored at Jefferson Lab. [10]

can be obtained from a measurement of the M_s dependent (e,e′p) coincidence cross section. Scattering off polarized deuterium can be described by replacing $\rho(p_m)$ by a spin projection (M_s) dependent momentum distribution $\rho_{M_s=0,\pm1}(p_m)$. The calculated $\rho_{M_s}(p_m)$ for the different M_s states is shown in Fig. 2. The angle θ_k is the angle between the momentum $\vec{p_m}$ and the quantization axis. Within the validity limits of PWIA, using a polarized target permits in principle the determination of the polarization dependent momentum distribution. If the nucleons really remain nucleons as described above, then the momentum distribution must have this dependence on the deuteron orientation.

2. Determination of ρ_{M_s}

The determination of the M_s-dependent D(e,e′p)n cross can be best shown in an idealized situation where the quantization axis (the target magnetic field $\vec{B}$) is along the momentum transfer $\vec{q}$ and is also the z-axis. Each target deuteron occupies one of the possible magnetic substates M_s. If $N_{-1}, N_0, N{+}1$ are the number of nuclei in the state $M_s = -1, 0, 1$, respectively, and $N = N_{-1} + N_0 + N_{+1}$ is the total number of target nuclei, one can define the normalized populations $n_i = N_i/N$ with $n_{-1} + n_0 + n_{+1} = 1$. For an unpolarized target $n_i = \frac{1}{3}$. The measured cross section σ can be now expressed in terms of the normalized populations and the M_s-dependent cross section σ_{M_s} as

$$\begin{aligned} \sigma &= n_{-1}\sigma_{-1} + n_0\sigma_0 + n_{+1}\sigma_{+1} \\ &= (n_{-1} + n_{-1})\sigma_{\pm 1} + n_0\sigma_0 \end{aligned} \tag{2}$$

where $\sigma_{-1} = \sigma_{+1} = \sigma_{\pm 1}$. From this one can express the unpolarized cross section as

$$\sigma_{unpol} = \frac{1}{3}\left(2\sigma_{\pm 1} + \sigma_0\right), \tag{3}$$

The polarization state of the spin 1 target can be described by its vector polarization

$$P_z = n_{+1} - n_{-1} \tag{4}$$

and the tensor polarization

$$P_{zz} = (n_{+1} + n_{-1} - 2n_0). \tag{5}$$

The target response is then expressed in terms of the target vector

$$A_z = \frac{\sigma_{+1} - \sigma_{-1}}{\sigma_{+1} + \sigma_{-1}} \tag{6}$$

and tensor analyzing powers.

$$\begin{aligned} A_{zz} &= \frac{(\sigma_{+1} - \sigma_0) + (\sigma_{-1} - \sigma_0)}{\sigma_{-1} + \sigma_0 + \sigma_{+1}} \\ &= \frac{2}{3}\frac{(\sigma_{\pm 1} - \sigma_0)}{\sigma_{unpol}} \end{aligned} \tag{7}$$

For the special case above $A_z = 0$ and the polarized cross section can be expressed in terms of the target polarization and the analyzing powers as follows:

$$\sigma_{pol} = \sigma_{unpol}(1 + \frac{1}{2}P_{zz}A_{zz}), \tag{8}$$

Therefore from the knowledge of the target polarization and the corresponding measured D(e,e′p)n cross section for the polarized target σ_{pol} and the measured cross section for an unpolarized target σ_{unpol} one can extract the cross sections $\sigma_{\pm 1}$ and σ_0 and from there the reduced cross sections which, if PWIA were valid, would correspond to the momentum distributions ρ_{M_s}. The relations are as follows:

$$\sigma_0 = \sigma_{unpol}\left(1 - \frac{2}{P_{zz}}\left(\frac{\sigma_{pol}}{\sigma_{unpol}} - 1\right)\right), \tag{9}$$

$$\sigma \pm 1 = \sigma_{unpol}\left(1 + \frac{1}{P_{zz}}\left(\frac{\sigma_{pol}}{\sigma_{unpol}} - 1\right)\right). \tag{10}$$

Journal of Physics: Conference Series **543** (2014) 012011 doi:10.1088/1742-6596/543/1/012011

3. Previous Experiments

Due to the challenges associated with building and operating a polarized deuterium target in an electron beam, very few electro-disintegration experiments have been carried out so far. Two experiments have been carried out at NIKEF. One measured the vector analyzing power A_V for missing momenta up to about 350 MeV/c [11]. The data show a large sensitivity to the deuteron D-state for missing momenta above 200 MeV/c and an increasing sensitivity to isobar configurations and FSI with increasing missing momentum. The other experiment [9] measured the tensor analyzing power A_T (A_{zz}) for missing momenta up to 150 MeV/c. The measured analyzing power is well reproduced by modern deuteron models and is mostly due to the D-state contribution even at the small missing momenta measured. The missing momentum range covered by the data from ref. [9] is comparable to the missing momentum range that could be covered at Jefferson Lab and was proposed in an earlier proposed Hall C experiment [10]. Clearly such a measurement would extend the data considerably and provide a complementary data set to the existing vector analyzing power data.

4. A Possible Experiment in Hall A at Jefferson Lab

In contrast to the original proposal a polarized deuterium target can now be operated in Hall A in combination with the two high resolution spectrometers. This makes it possible to increase the beam energy to 4.4 GeV and reduce the electron scattering angle to 7°. This small electron scattering angle increases the cross section by about a factor of two compared to the original proposal and makes it also possible to increase the momentum transfer thus reducing contributions due to FSI. A possible set of kinematics is shown in Table 1.

p_m MeV/c	ω MeV	q MeV/c	Q^2 (GeV/c)^2	x_B	θ_q deg	p_f MeV/c
200.0	325.0	610.7	0.27	0.44	54.4	810.7
250.0	394.6	646.9	0.26	0.35	49.0	896.8
300.0	481.2	699.0	0.26	0.28	43.1	999.0
350.0	591.5	774.4	0.25	0.23	36.8	1124.4
400.0	736.3	884.6	0.24	0.17	30.3	1284.6
450.0	935.2	1050.0	0.23	0.13	23.7	1499.7

Table 1. Kinematic settings for an incident energy $E_i = 4.4$ GeV and and electron scattering angle $\theta_e = 7°$. θ_q is the angle of the momentum transfer $\vec{q}$ with respect to the beam which

The momentum transfer corresponds closely to the one of the A_V experiments [11]. The high final electron momentum ($3 < p_e < 4$ GeV) would reduce the effect of the target magnetic field considerably. The proton momenta are relatively low but as long as the protons are emitted along the magnetic field the effect of the strong target field as used in the UVA target [12] is also reduced. The problem of a coincidence experiment with a polarized target is that the magnet itself imposes many restrictions with respect to the accessible particle directions. It is expected that not all kinematic settings will be able to be measured in the ideal parallel kinematics but that the magnetic field and the momentum transfer will be at an angle with respect to each other. In this case the simple relationships between measured polarized cross section and the magnetic substate populations is complicated and other terms of polarization and analyzing powers need to be included. These additional terms can be viewed as corrections and can be calculated. The calculated cross sections for $M_s = 0$ and $M_S = \pm 1$ for the kinematic settings of table 1 are shown in Fig. 3.

The solid lines with the diamond symbols indicate the cross sections for the $M_s = 0$ state and the dash-dot lines with the square symbols are for the $M_s = \pm 1$ states. All calculations

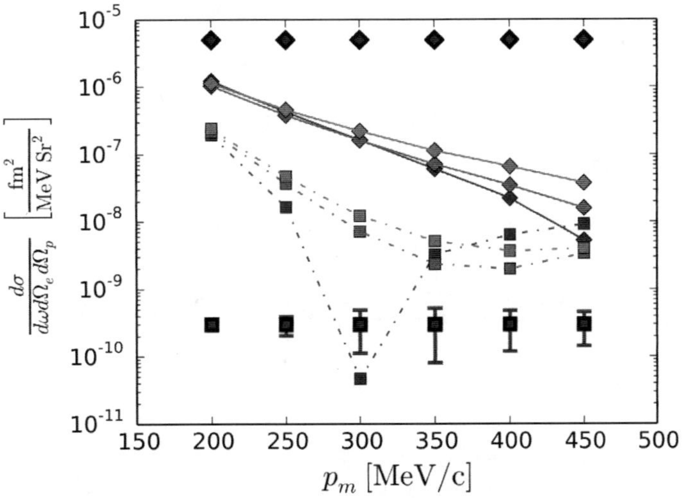

Figure 3. Calculated cross sections for the kinematic setting of Table 1. Solid lines and diamonds indicate $M_s = 0$ cross sections. Dash-dot lines and square symbols are for $M_s = \pm 1$ cross sections. PWIA (blue lines), FSI(green lines) and FSI+MEC+IC (red Lines) are from a calculation by R. Schiavilla using the argonne V18 [13] potential. The top row of diamond symbols indicated the expected uncertainties in σ_0 and the bottom row of square symbols with error bars show the anticipated error in $\sigma_{\pm 1}$ for an experiment of about 30 days.

are from R. Schiavilla using the Argonne V18 potential [13]. The blue lines are from a PWIA calculation while the green lines include FSI and the red ones in addition include MEC and IC. The top row of diamonds indicate an estimated uncertainty for the $M_S = 0$ cross sections (of the order of 6 - 7%) and the bottom row of squares with error bars show the anticipated uncertainty in the $M_s = \pm 1$ cross section. The calculated tensor analyzing power together with the anticipated uncertainty is shown in Fig. 4 where the colors have the same meaning as in Fig. 3. A_{zz} seems to be very little affected by MEC and IC for missing momenta below 400 MeV/c but this changes for missing moment of 450MeV/c and above. FSI seems to contribute to A_{zz} at the level of 15 - 20% below a missing momentum of 350 MeV/c. Above this missing momentum, FSI contributions are predicted to increase rapidly leading to a change of sign in A_{zz} at p_m= 450 MeV/c. The error estimates are based on a beam current of 80 nA and a tensor polarization of 20%. In addition it was assumed that the unpolarized cross section has been measured with a statistical precision of about 1%. It was also assumed that the tensor polarization has been measured with a relative precision of 7%.

5. Summary and Conclusion

Exclusive electro-disintegration of tensor polarized deuterium at low momentum transfers of about 0.25 $(\text{GeV/c})^2$ and up to missing momenta of 450 MeV/c will provide new, detailed information on the spin structure of the deuteron in a kinematic regime where the D-state starts to dominate the ground state. While for missing momenta below 350 MeV/c, FSI and non-nucleonic degrees of freedom are expected to be relatively small their contributions are predicted raise quickly for missing momenta above 400 MeV/c. These data would also provide a stringent test of the quality of modern non-relativistic models (including relativistic corrections). Jefferson Lab is currently the only facility in the world where such a measurement could be carried out in

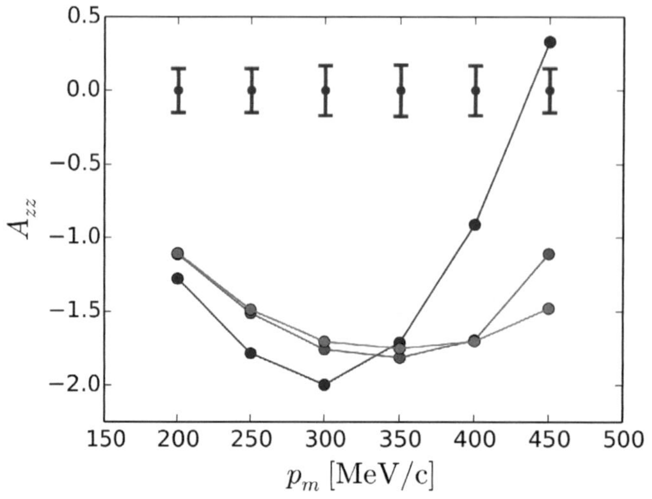

Figure 4. Calculated tensor analyzing power for the kinematic setting of Table 1 where the magnetic field $\vec{B}$ is parallel to $\vec{q}$ and the ejected proton is parallel to $\vec{q}$ as well. PWIA (blue lines), FSI(green lines) and FSI+MEC+IC (red Lines) are from a calculation by R. Schiavilla using the argonne V18 potential. The error bars show the estimated uncertainty from a potential experiment of about 30 days.

combination with a dynamically polarized ND_3 [12] or potentially a LiD [14] target. In order to make a more accurate assessment of the needs of such an experiment the details on the target system to be used need to be further specified, such as the target material and the achievable polarization. One also needs to study background contributions of the other materials besides deuterium and a way of correcting or subtracting these contributions from the measured cross section.

References

[1] Subedi R *et al.* 2008 *Science* **320** 1476
[2] Shneor R *et al.* 2007 *Phys. Rev. Lett.* **99** 07251
[3] Piasetzky E, Sargsian M, Frankfurt L, Strikman M and Watson J W 2006 *Phys. Rev. Lett.* **97** 162504
[4] Schiavilla R, Wiringa R B, Pieper S C and Carlson J 2007 *Phys. Rev. Lett.* **98** 132501
[5] Sargsian M M, Abrahamyan T V, Strikman M I and Frankfurt L L 2005 *Phys. Rev.* **C 71** 044615
[6] Alvioli M, degli Atti C C and Morita H 2008 *Phys. Rev. Lett.* **100** 162503
[7] Blomqvist K *et al.* 1998 *Phys. Lett. B* **424** 33
[8] Forest J *et al.* 1996 *Phys. Rev. C* **54** 646
[9] Zhou Z L *et al.* 1999 *Phys. Rev. Lett.* **82** 687
[10] Anklin H and Boeglin W U 1998 *Jefferson Lab Proposal E-97-102*
[11] Passchier I *et al.* 2002 *Phys. Rev. Lett.* **88** 102302
[12] Pierce J *et al.* 2014 *Nucl. Instrum. and Meth. A* **738** 54
[13] Schiavilla R 2014 *priv. comm.*
[14] Bültmann S *et al.* 1999 *Nucl. Instrum. and Meth. A* **425** 23

Journal of Physics: Conference Series **543** (2014) 012012 doi:10.1088/1742-6596/543/1/012012

Systematic Errors in measurement of b_1

S A Wood

Thomas Jefferson National Accelerator Facility, Newport News, Virginia 23606, USA

E-mail: saw@jlab.org

Abstract. A class of spin observables can be obtained from the relative difference of or asymmetry between cross sections of different spin states of beam or target particles. Such observables have the advantage that the normalization factors needed to calculate absolute cross sections from yields often divide out or cancel to a large degree in constructing asymmetries. However, normalization factors can change with time, giving different normalization factors for different target or beam spin states, leading to systematic errors in asymmetries in addition to those determined from statistics. Rapidly flipping spin orientation, such as what is routinely done with polarized beams, can significantly reduce the impact of these normalization fluctuations and drifts. Target spin orientations typically require minutes to hours to change, versus fractions of a second for beams, making systematic errors for observables based on target spin flips more difficult to control. Such systematic errors from normalization drifts are discussed in the context of the proposed measurement of the deuteron b_1 structure function at Jefferson Lab.

1. Introduction

The deuteron tensor structure function b_1 has been measured at intermediate x with HERMES [1]. An experiment at JLab (E12-13-11 [2]) seeks to improve upon this measurement, taking advantage of the substantially higher luminosity that can be achieved with solid targets. b_1 can be related to the tensor asymmetry A_{zz} by

$$b_1 = -\frac{3}{2} F_1^d A_{zz}. \tag{1}$$

This asymmetry is obtained from the difference of inclusive electron scattering cross sections on deuterium in tensor polarized (with polarization P_{zz}) and unpolarized states.

$$A_{zz} = \frac{2}{P_{zz}} \frac{(\sigma_p - \sigma_u)}{\sigma_u} = \frac{2}{f P_{zz}} \frac{(Y_p - Y_u)}{Y_u} = \frac{2}{f P_{zz}} A_{zz}^{\text{raw}}. \tag{2}$$

As the proposed JLab measurement uses an ND_3 target rather a pure deuteron target, the experimentally observed asymmetry must be divided by a "dilution factor" f which represents the fraction of the total yield due to scattering on deuterium. For an ND_3 target, which includes impurities such as helium and target wall materials, the dilution factor is

$$f = \frac{N_D \sigma_D}{N_N \sigma_N + N_D \sigma_D + \sum A N_A \sigma_A}. \tag{3}$$

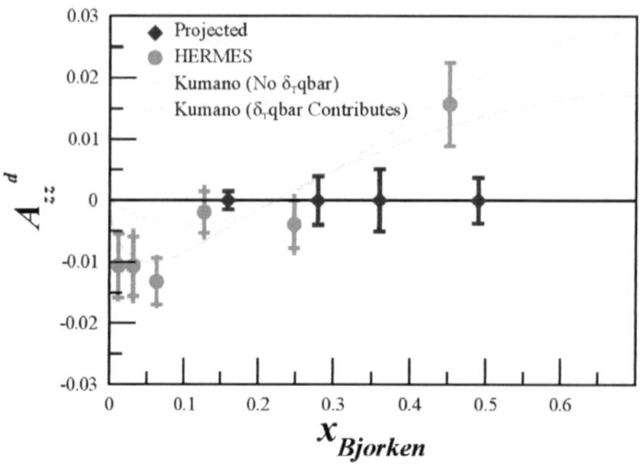

Figure 1. Projected A_{zz} statistical errors for the JLab b_1 experiment.

The dilution factor does vary as function of scattered electron energy (or x), particularly at kinematics where nucleon resonances are prominent. However, for the purpose of discussing systematic errors, it is reasonable to assume that $\sigma_A/\sigma_D = 14/2$ and that the $\sum$ term (accounting for other materials such as helium and target wall material) can be neglected. These assumptions then give a dilution factor of $f \approx 3 \times 2/(14 + 3 \times 2) = 0.3$.

2. Measuring beam asymmetries

The expression for the A_{zz} asymmetry is similar to the expression for asymmetry based on flipping beam helicity,

$$A = \frac{1}{P_b} \frac{(\sigma_+ - \sigma_-)}{(\sigma_+ + \sigma_-)}, \tag{4}$$

where P_b is the polarization of the beam and σ_+ and σ_- are the (doubly differential) cross sections of the process of interest for two opposite beam polarization directions. Measurement of such asymmetries are attractive, because in the method of converting scattering yields to cross sections,

$$\frac{d\sigma(E_b, \theta, E)}{dEd\Omega} \leftarrow N(E_b, \theta, E) \frac{1}{\Delta\Omega} \frac{1}{\Delta E} \frac{e}{Q} \frac{A}{t\rho N_A} \frac{1}{\epsilon}, \tag{5}$$

the various normalization factors and the kinematics (beam energy, scattered particle angle, scattered particle energy) are naively identical (even if not accurately known) for the two polarization states. (Q is accumulated beam charge passing through a target, t is target thickness, ϵ is the overall detector efficiency, while $\Delta\Omega$ and ΔE are the solid angle and energy acceptances of the detector.) If the normalization factors and kinematics are indeed identical for the two spin directions, then errors in the asymmetry are only from statistics and any errors in the beam polarization P_b.

All of these normalization and kinematic quantities can and do drift or fluctuate with time to varying degrees and can even be correlated with spin direction. Effects that lead to errors in the asymmetry are generally referred to as false asymmetries or systematic errors. In many classes of measurements, the asymmetry is large enough that these effects are negligible. But in the push to access fundamental nucleon properties with parity violating electron scattering, asymmetries smaller than 1×10^{-6} are being routinely targeted. Some examples of effects that can give false asymmetries at this level are:

- Density fluctuations in a liquid target that exceed statistical fluctuations.

- Helicity correlated beam properties so that σ_+ and σ_- are measured at different beam energies or kinematics
- Fluctuations in the calibration of beam current measuring devices

While the systematic errors from any of these uncertainties can easily swamp the asymmetry, good planing and experiment design can overcome these issues. Techniques employed in experiments such as HAPPEX, G0 [3], and Qweak [4] include rapid helicity reversal, attention to target design, beam diagnostics, beam property feed back systems and others.

3. Fixed Target Asymmetries

While fundamentally the same, measuring asymmetries between different fixed target spin states differs in one important manner, namely that the time required to change the magnitude or direction of target spin ranges from minutes to hours. Experimental parameters and conditions that may be stable at the one second level may fluctuate at longer time scales due to environmental effects including temperature, barometric pressure and power conditions.

Potential contributors to systematic errors include:

- Beam
  - **Current measurement** Instability of the calibration of beam charge measurement devices.
  - **Beam energy** Relative yield changes from the strong Q^2 dependence of electron scattering cross sections can be comparable to raw target asymmetries.
  - **Beam direction** A change in beam direction changes the scattering angle, changing yield.
- Target
  - **Thickness stability** The ND_3 is composed of beads, which allows liquid helium to flow around the target material. With time, the position of these beams can shift, causing the amount of material seen by the beam to change.
  - **Beam motion** For a target of non-uniform thickness, such as ND_3, a change in beam position will change the target thickness seen by the beam.
- Detection
  - **Detectors** Drifts in photomultiplier tube (PMT) gain can change the number events above a discriminator threshold or that pass particle identification (PID) cuts. Atmospheric pressure and temperature changes can affect drift chamber efficiency.
  - **Spectrometer field** A drift in spectrometer analyzing field will change yields both due to cross section dependence on scattered electron energy and due to the changed momentum acceptance.
  - **Target holding field** For polarized targets with large holding fields, drifts in this field can change the kinematics (scattering angle and energy) of particles reaching the detectors.

The extent to which any of the above contribute to systematic errors in a target asymmetry depends upon the desired accuracy in the asymmetry measurement, the time-scale at which the target spin state can be cycled, and the degree to which the stability of the above factors can be controlled or monitored.

4. Systematic errors in b_1 measurement

Measurement of b_1 will push the requirement to understand and control instabilities in normalization factors and experimental conditions. The existing measurements, which used a polarized deuterium gas jet target, show a typical physics asymmetry A_{zz} of $\sim$0.01 (Fig. 1). The JLab b_1 experiment proposes to measure this asymmetry with an error $\Delta A_{zz}\sim$0.005 or better, taking advantage of the higher available luminosity. The trade-off with the increased

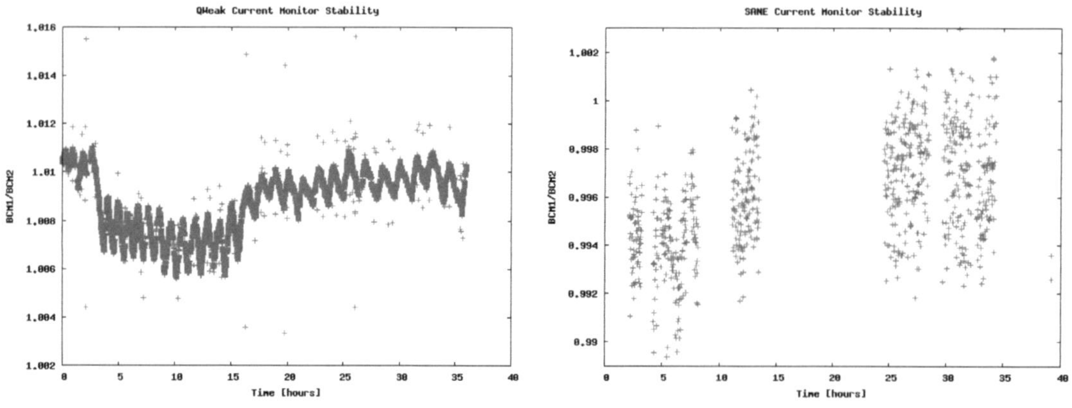

Figure 2. The left figure shows the ratio of the two standard Hall C beam current monitors over a 40 hour period during the Qweak experiment with a beam current of $\sim$180 μA. The right figure shows the ratio of these same current monitors over a 40 hours period during the SANE experiment with a beam current of $\sim$100 nA. In both cases, the relative stability of the current monitors is worse than 1×10^{-3}, showing oscillations with a period of about an hour and longer term drifts on the scale of a day.

luminosity is that the scattering yield from deuterium in a polarized ND_3 target is diluted by the nitrogen and the deuterium tensor polarization is less than what is available from gas jet targets. With the dilution and assuming that $P_{zz} = 0.20$, Eq. 2 implies that a raw asymmetry precision of better than $\Delta A_{zz}^{\mathrm{raw}} = \frac{fP_{zz}}{2}\Delta A_{zz} = \frac{0.3\times 0.2}{2}0.005 = 1.5 \times 10^{-4}$ is required.

The polarized ND_3 target takes about an hour to achieve full polarization from an unpolarized state. Therefore, it is impractical to change the target polarization state more than 2 or 3 times per day without losing significant beam time. This both subjects the measurement to diurnal drifts and limits the number of times that the polarization state can be changed in order to average out fluctuations.

An example of an issue which, if unaddressed, would render the b_1 measurement impossible, is the stability of the beam current monitors (BCMs) used to measure charge incident on the target. Figure 2 shows the ratio of current measured by two different BCMs in Hall C at JLab. The relative calibration of these two BCMs shows instabilities greater than 1×10^{-3} at time scales of both an hour and a day. It is important to note that the BCMs and the data acquisition have not been optimized for target asymmetry experiments, but rather only to have a stability better than 0.5%, which is sufficient for cross section measurements.

Some of the more subtle potential contributions to systematic errors in A_{zz} are instabilities in beam energy, beam direction and magnetic field of the detection spectrometer. The contributions of each of these depends on the sensitivity of the yield to each of these parameters and the precision to which these parameters can be controlled or measured over times scales of hours to days. Table 1 shows the sensitivity of the deuterium electron scattering cross section (from the Bosted/Christy [5] cross section fit) to these parameters. These sensitivities indicate that in order to keep the systematic errors on A_{zz} less than 1×10^{-4}, both the beam energy and the scattered electron energy must be controlled to better than 1×10^{-4}. Similarly, because a change in beam direction is a change in scattering angle, the beam direction must in some cases be held stable to better than 5×10^{-6} radians. (Note: There are no strong requirements on energy/angular spread or on the accuracy of the absolute determination of these parameters.)

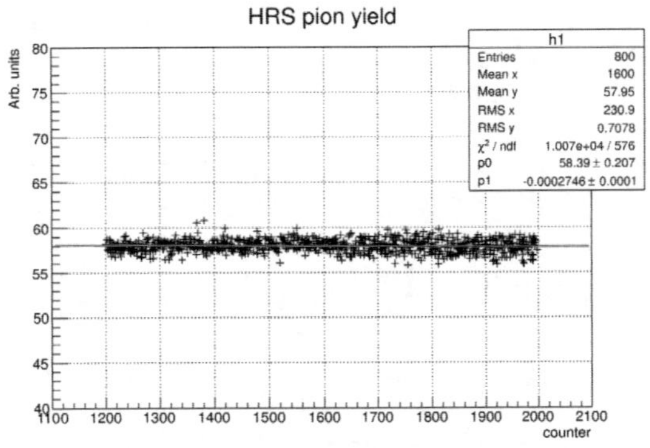

Figure 3. The pion yield in the HRS spectrometer over a 15 day period during Hall A Transversity Experiment, E06-010 [6]. The primary measurement during this experiment was $(e, e'\pi)$ coincidence events. Higher rate pion singles were monitored during experiment, providing a check on the stability of the product of beam current calibration, target thickness and detector efficiency. A straight line fit to the yield (pion counts normalized by beam charge and dead time) shows a drift of 0.38% over 15 days, or a drift of 2.5×10^{-4} per day.

5. Controlling Systematic Errors

The Hall A Transversity experiment [6], E06-010, demonstrated that the stability requirements of the b_1 experiment may be achievable. This experiment measured target asymmetries of the $(e, e'\pi)$ reaction on a polarized ^{3}He target. As this was a coincidence measurement, the high singles rate in one of the spectrometers (from the inclusive reaction $^3\text{He}(e, \pi)$), could be used as a stability monitor. Figure 3 shows the singles rate normalized only by beam current and deadtime, providing a monitor of the combined stability of BCM calibration, target thickness, detector efficiency, spectrometer optics and beam properties. The long term trend of this normalized yield is a drift of 2.5×10^{-4} per day, which is near the stability requirement of b_1 measurements.

As the b_1 experiment is a singles measurement, there is no higher rate background reaction (with a null or known target asymmetry) that can be used as a simultaneous measure of the combined stability of beam, target and detector parameters and calibrations. Further study of

Table 1. Estimates of the sensitivity of the $d(e, e')$ cross section and yield to variations in beam energy (E_b), scattered electron energy (E_o) and scattering angle θ for the kinematics of the b_1 proposal, calculated from the Bosted/Christy fits to electron-deuteron scattering cross sections [5]. The fourth column, the yield sensitivity to spectrometer field changes, includes the change in the momentum acceptance from the change in spectrometer momentum. The last column, the sensitivity to scattering angle, is in units of inverse radians.

x	E_b $\frac{d\sigma}{dE_b}\frac{E_b}{\sigma}$	E_o $\frac{d\sigma}{dE_o}\frac{E_o}{\sigma}$	E_o $\frac{dY}{dE_o}\frac{E_o}{Y}$	θ $\frac{d\sigma}{d\theta}\frac{1}{\sigma}$
0.16	-2.03	0.28	1.28	-9.6
0.30	-1.08	-1.07	-0.07	-14.0
0.45	-0.73	-2.22	-1.22	-18.7
0.55	-0.37	-4.06	-3.06	-20.9

previous and future experiments which have long periods of data acquired at fixed kinematic settings can be useful. These studies can help to determine what experimental parameters are the most unstable and thus require the most effort to develop techniques to mitigate the resultant systematic errors.

In order to control the systematic errors on the measurement of b_1, a number of techniques will need to be developed to monitor the stability of the various experimental parameters that can contribute to these errors. Some possible measures include:

- Take periodic measurements of yield using higher currents on a solid target to provide a periodic check of the stability of the combination of beam current calibration and detector efficiency. Since inclusive nuclear cross sections have similar dependencies on incoming energy, outgoing energy and scattering angle as deuterium cross sections, these measurements also fold in the stability of those quantities. These high yield measurements could be made during periods in which the polarized target is being repolarized.
- Improve Beam Current Monitors. The stability of the standard JLab BCMs can be improved by providing better temperature stabilization of the BCM cavities, stabilizing the temperature of analog cables and minimizing the length of those cables. Averaging multiple BCMs, with independent temerature stabilization can further improve the stability of current measurement.
- Independent current measurement techniques should be explored. As the ND_3 target limits the beam current to about 100 nA, a Faraday Cup can continuously monitor current and provide a stability check on other beam current measurements.
- Luminosity monitors, small integrating detectors placed at small scattering angles downstream of the target, provide a relative monitor of the product of beam current and target thickness. While luminsoity monitors have a high statistical precesion, care must be taken to demonstrate that the luminosity detectors have an long term efficiency stability that is better than the stability requirement of the experiment. However, even in the absence of this stability, such detectors can be used to detect sudden changes in target thickness from shifting of the ND_3 beads or a change in beam position on the target.
- The gain of PMTs should be stabilized as much as possible by keeping stable temperature and high voltage. Several PMTs could be instrumented with good temperature and voltage probes in order to establish that these quantities are stable. The thresholds on discriminators should be set as low as practical and PID cuts in analysis should be set as wide as possible to minimize the sensitivity to drifts in PMT gain.
- Existing techniques to stabilize beam energy and orbit should be extended. Techniques that focus on long term stability, including carfeul monitoring of transport dipole magnetic fields and a Synchrotron Light Interferometer, have been developed for low rate spectroscopy measurements such as the hypernuclear program [7].

In the above, the emphasis needs to be on stability and the accurate monitoring of changes. Requirements on the absolute determination of experimental parameters such as beam energy, current, target thickness, solid angle, etc. are similar or more relaxed than the requirements of typical spectrometer experiments. For some parameters, such as detector efficiencies, it may not be possible to monitor their stability with sufficient accuracy during the measurement. For such cases it may be necessary to use other higher rate reactions to study the stability of the parameter and its sensitivity to measurable environmental conditions such as temperature and air pressure.

6. Conclusion

Measurements of quantities such as b_1 that are derived from target asymmetries present unique challenges when measuring raw asymmetries with an accuracy of the order of 1×10^{-4}.

Journal of Physics: Conference Series **543** (2014) 012012 doi:10.1088/1742-6596/543/1/012012

Comprehensive identification of all potential sources of systematic errors, some of which are described here, will be required. New instrumentation and techniques will need to be developed well in advance of experiments that seek to measure such asymmetries. Use of these techniques in other experiments with less stringent requirements will help to gain experience and establish the feasibility of controlling and monitoring experimental conditions over long time periods.

Acknowledgments

The author would like to acknowledge the JLab b_1 collaboration for discussions of the details of the proposed measurement, Buddhini Waidyawansa for data on BCM stability during Qweak, Kalyan Allada for data on long term yield stability and Dave Mack for many suggestions on mitigating systematic errors. This work was supported by DOE Contract DE-AC05-06OR23177 under wich Jefferson Science Associates, LLC operates Jefferson Lab.

References

[1] Airapetian A *et al* 2005 *Phys. Rev. Lett.* **95** 242001
[2] Slifer K *et al* 2013 E12-13-011: The deuteron tensor structure function b_1 *Proposal to Jefferson Lab PAC40*
[3] Androic D *et al* 2011 *Nucl. Instr. Methods A* **646** 59
[4] Androic D *et al* 2013 *Phys. Rev. Lett.* **111** 141803
[5] Bosted P E and Christy M E 2008 *Phys. Rev. C* **77** 065206
[6] Allada K *et al* 2014 *Phys. Rev. C* **89** 042201
[7] Tang L *et al* 2014 *arXiv*:1406.2353

Polarized target options for deuteron tensor structure function studies

O A Rondon-Aramayo

Institute for Nuclear and Particle Physics, U. of Virginia, Charlottesville, VA

E-mail: or@virginia.edu

Abstract. Studies of spin-dependent observables generally rely on calculating the difference between scattering yields for opposite spin states of the beam helicity or the target polarization or both. While the beam helicity can be changed at rates of up to many times per second, target polarization flips involve times on the order of hours or longer. To measure observables that depend only on the differences for opposite target polarizations with minimal systematic effects caused by changes in the beam, target, and detector systems, special methods need to be followed. One option is to use targets which have two or more sections in the beam path, each with materials polarized independently. While this approach does not remove all sources of systematic effects, it can be combined with frequent alternation of spin states to improve the control of time dependent and configuration uncertainties.

1. Introduction

A general expression for spin dependent electromagnetic scattering cross-sections looks like [1, 2]

$$\sigma = \sigma^U(1 + \cancel{hA^h} + P_V A_V + P_T A_T + h(P_V A_V^h + P_T A_T^h)) \tag{1}$$

where σ^U is the spin independent cross section, h is the beam helicity, P_V and P_T are the vector and tensor target polarizations, respectively, A_V and A_T are single spin vector and tensor target asymmetries and A_V^h and A_T^h are the corresponding double spin ones. The out-of-plane beam A_h is crossed out because it is absent in inclusive scattering. Moreover, A_T and A_T^h are zero for spin $< 1\hbar$.

The target asymmetries depend on opposite target polarizations, which are hard to flip, unlike the beam helicity. The experimenter is then faced with two options to flip $P_{t(arget)}$:

- use a single cell target with periods of alternating P_t
- use a target with multiple cells, each with independent P_t (which could also be zero.)

An example of such observable is the deuteron tensor structure function (SF) b_1^d. The cross section for unpolarized electrons scattering on longitudinally polarized deuterons [3] as a function of the Bjorken scaling variable x and the four-momentum transfer Q^2 is

$$\sigma^{\parallel}(x, Q^2) = Kx(F_1^d(x, Q^2) - \frac{P_{zz}}{3} b_1^d(x, Q^2)) \tag{2}$$

where $K(y, E, Q^2)$ is a kinematical factor which depends on the beam energy E and the fractional energy transfer $y = \nu/E$, $\nu = E - E'$ is the difference between the beam and the scattered energy.

For $P_{zz} = 0$ one obtains the unpolarized cross section $\sigma^U = KxF_1^d$. The tensor SF can then be isolated from the cross section difference

$$\sigma^U - \sigma^{\|} = Kx\frac{P_{zz}}{3}b_1^d.$$

b_1 can be measured by taking data simultaneously on two (or more) unpolarized and polarized in-line cells: the yields from all cells share the beam charge and detector efficiency. Moreover, for multiple data taking periods with alternating cell polarizations, one can form yield differences for both beam upstream–downstream cells, and same cell differences for different polarization periods.

2. Multiple cell polarized targets

The concept of multiple cell polarized targets is not novel. It was introduced introduced by the EMC experiment (CERN NA2') to deal with the single sign of the polarization of muon beams (always negative). It consisted of upstream and downstream cells [4], with separate microwave feeds, as indicated on Fig. 1. A similar configuration was used in subsequent targets, like those used in SMC [5], see Fig. 2, and COMPASS [6], see Fig. 3.

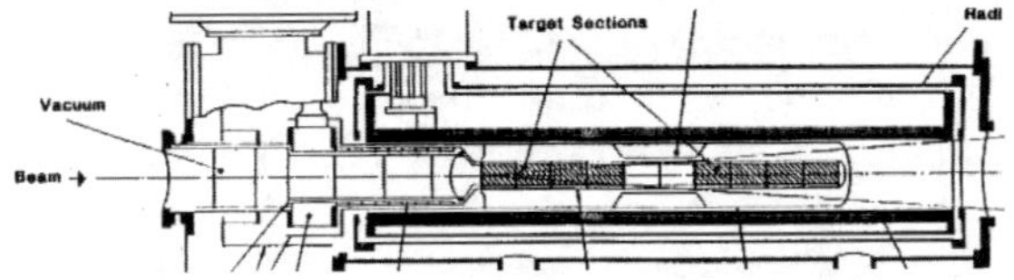

Figure 1. EMC dual cell target.

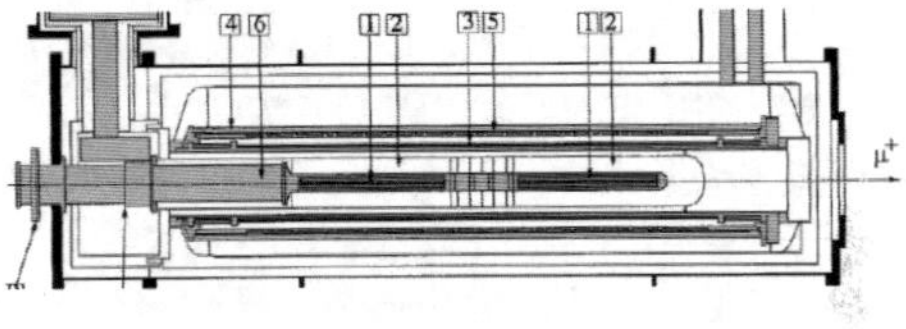

Figure 2. SMC dual cell target.

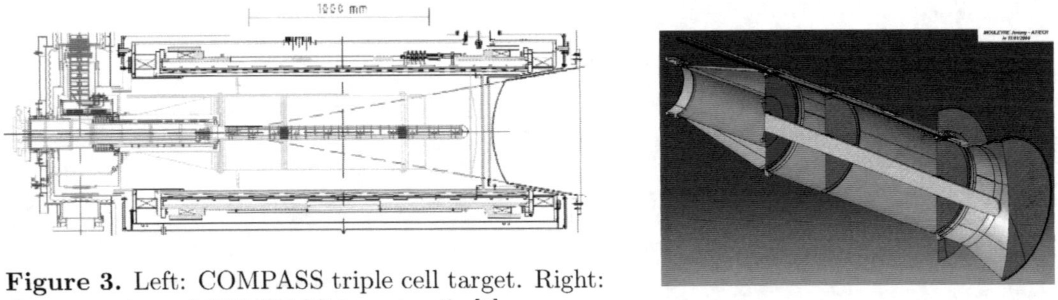

Figure 3. Left: COMPASS triple cell target. Right: Cutaway view of COMPASS target cells [7].

Multiple cell targets have favorable and unfavorable features. Among the pros one can count

- can compensate for few or no polarization reversals by taking data simultaneously on two or more polarization states
- use favorable cylindrical symmetry for simple, intense solenoidal magnetic fields ≥ 6.5T
- increased data rate, allowing lower beam currents than single cell systems, which results in fewer anneals and material changes
- reduced or eliminated time dependent systematic effects.

The cons include

- more complicated systems for refrigeration, microwaves feeding and isolation

- increased radiation thickness increases external radiative corrections
- event loss from acceptance matching cuts
- cell dependent packing fractions.

3. A more detailed look

In order to evaluate the possibility of using multiple cell targets in experiments at JLab, a more detailed look at their features is needed.

The most important component of multiple cell targets are the coils. Their geometry and configuration determine critical aspects, such as angular acceptance and achievable polarization levels. The CERN targets used a solenoidal geometry with a main coil with a length L to radius R aspect ratio $L/R = 20/3.3$. This ratio determines in part the angular apertures for scattered particles and the need for additional trim and compensation coils to achieve the field homogeneity required for dynamic nuclear polarization - DNP, over a region of significant size where the polarizable material is located.

The solenoid magnet of the COMPASS target used coils first built by Saclay [6], shown in Fig. 3, which were later upgraded to ones built by Oxford Instruments [8], shown in Fig. 4. The magnet in both cases consisted of a 2.7 T solenoid plus 0.5 T dipole for transverse field. Fig. 5 illustrates the coil configuration for the Saclay magnet. The field had a better than 10^{-4} uniformity and it initially accommodated two 30 mm diameter cylindrical cells 60 cm long each with a 5 cm spacing in between. The cells were later upgraded to 40 mm diameter ones, with three sections: upstream + downstream, 30 + 30 cm, and a central 60 cm long one, shown on Fig. 3.

With the Oxford magnet the system had a $\pm 14°$ downstream acceptance, while the Saclay magnet allowed only ± 75 mr, or less than $5°$.

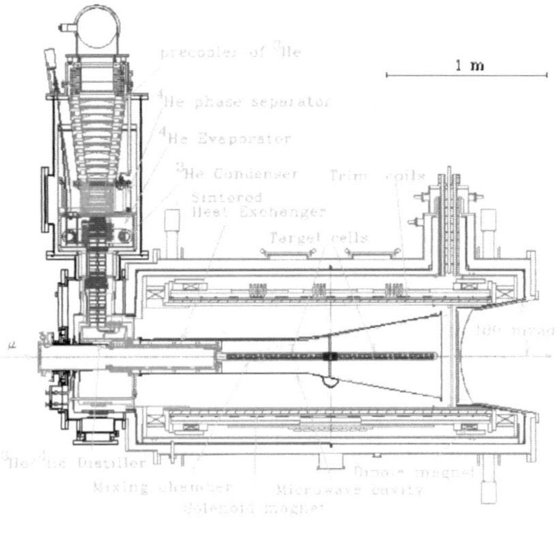

Figure 4. Diagram of the COMPASS target with the Oxford Instruments magnet [8].

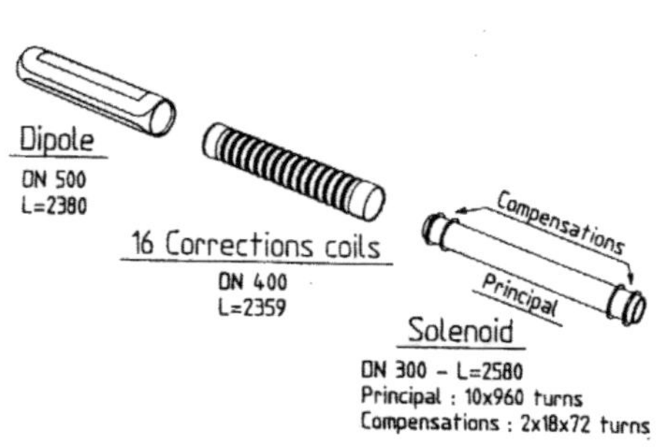

Figure 5. Exploded view diagram of the Saclay coils [9].

Given the excellent characteristics of the COMPASS Oxford magnet, it serves as a guide for designing a similar system for JLab polarized target experiments that would benefit from data simultaneously taken on multiple cells. Some criteria on magnet design for a multi-cell target for high intensity electron beams would be

- Solenoid for longitudinal field, including trim and compensation coils to achieve equal or better than 10^{-4} uniformity over the length of the cups. This could be done for JLab with a 1/5 scale copy of the Oxford magnet, with a slightly better aspect ratio of $L/R = 20/3$
- A uniform region of ± 6.5 cm, or a tenth of COMPASS's ± 65 cm, which would help in reducing the number of required correction coils.
- An optional dipole for transverse field.

For a 1/5 scale coil relative to the COMPASS magnet, the ratio $BL(\text{JLab})/BL(\text{COMPASS}) = 6.75\text{ T} \times 40\text{ cm}/(2.7\text{ T} \times 200\text{ cm}) = 0.5$ means that the NI ampere turns for a 6.75 T JLab solenoid would be 1/2 that of the COMPASS coil.

An example of a possible coil design is illustrated in Fig. 6, consisting of a 40 cm long, 6 cm radius solenoid. The uniform field region would extend over ± 6.5 or 7 cm, to fit two cells, each about 4 cm long, spaced by up to 5 cm to obtain clean vertex separations. The minimum downstream acceptances for this setup would be about 10.7° for the 4 cm cups spaced 5 cm. The acceptance could exceed 11.8° for 3 cm cups spaced 2 cm displaced downstream along the coil center. The optimal configuration would be determined after detailed modeling of coil and cell characteristics.

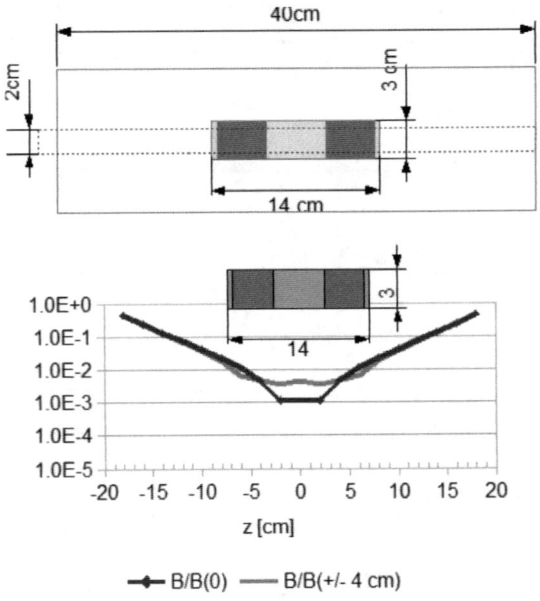

Figure 6. Possible solenoid coil design for JLab. The dashed lines represent the 1 cm radius envelope of the rastered beam. The curves represent the ratio of the pure solenoidal field along the axis (blue online, dark curve with diamonds) to the field at the center, and at ± 4 cm from the center (red online, light smooth curve).

4. Using multiple cell targets

Special considerations need to be applied when designing experiments and analyzing data taken with multiple cell targets. The most important one probably is to have good vertex reconstruction to identify data from individual cells. Clean separation with small loss of events from the ends of the cells is a must. This can be achieved by allowing adequate spacing between cells, at the possible price of complicating the matching of acceptances. Fig. 7 illustrates the vertex separation after cuts obtained with the triple cell COMPASS target [10].

For the case of the JLab spectrometers, figure 8 top panel shows the results of simulated vertex reconstruction with the Hall C High Momentum spectrometer - HMS at 14°, for an arrangement of two 4 cm long cells filled with solid ammonia and liquid He, located upstream

Journal of Physics: Conference Series **543** (2014) 012013 doi:10.1088/1742-6596/543/1/012013

and downstream of the target center and separated by 5 cm filled with LHe only, with additional 0.5 cm LHe at their extreme ends, like the example shown on Fig. 6. The bottom panel of the same figure shows the simulation for the Super HMS at 12°, this time with no LHe between the cells, but 0.5 cm LHe at all four ends. 11 GeV electrons were incident in both cases. No cuts on the scattering polar and azimuthal angles were applied. The upstream and downstream distributions can be matched to better than 1% applying angle acceptance cuts, with a tolerable yield loss of about 60%. The simulated data were prepared by Narbe Kalantarians.

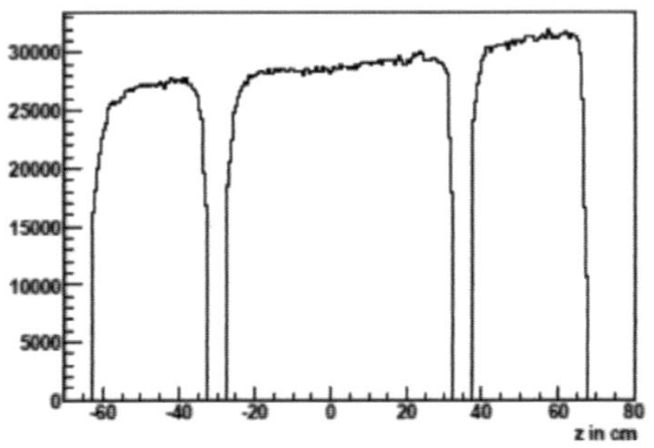

Figure 7. Vertex reconstruction, after cuts, of data taken with the triple cell COMPASS target.

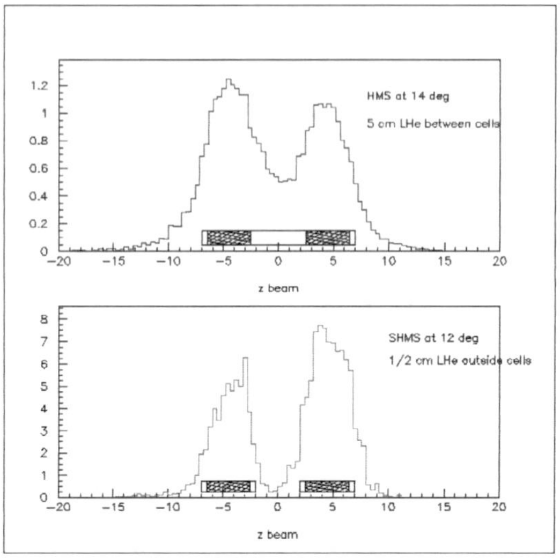

Figure 8. Top: vertex reconstruction of simulated HMS data. Bottom: SHMS vertex reconstruction.

As mentioned above, many types of yield differences can be formed with multi-cell target configurations, which are not possible with single cell systems. Moreover, each type is subjected to different systematic effects.

Time dependent effects, such as changing beam charge calibrations and detector efficiency, which affect single cell systems, aren't present for multiple cell ones. On the other hand, multi-cell setups have to deal with cell dependent acceptances $\mathcal{A} = \Delta E'\Delta\Omega$. In both cases, the total amount of polarizable material, or packing fraction, is also a source of systematic differences, time dependent ones for both cases and cell dependent ones for multiple cells.

For multiple cells it is important to optimize the length vs the size of the radiative corrections by balancing faster counting vs systematic errors. For example, for 3 + 3 cm cells with 60% packing fractions, each $4.6\%X_0$ long, the increase in external radiative corrections could be more than compensated by the doubling of the total Born rate compared with the single cups used in past polarized target experiments in JLab's Hall C, such as G_E^n (E93-028), RSS (E01-006) and SANE (E07-003).

But the main advantage of multi-cell targets is that many types of differences between subsets of data can be formed. Examples of possible differences are shown on the table of Fig. 9. COMPASS used a similar labeling scheme for their triple cell, even subdividing the large central cell into two virtual subcells [11].

	Cell		Differences		Changes
Period	*U*	*D*	Different cells	Same cell	
1	U_0	D_p	$\Delta_1^d = U_{01} - D_{p1}$		$\mathcal{A}_{U,D}$, $pf_{U,D}$
2	U_p	D_0	$\Delta_2^d = U_{p2} - D_{02}$	$\Delta_U^s = U_{01} - U_{p2}$, $\Delta_D^s = D_{p1} - D_{02}$	$Q_{1\text{-}2}$, $\varepsilon_{1\text{-}2}$, $pf_{1,2}(U), pf_{1,2}(D)$
3	U_p	D_0			
4	U_0	D_p			
			$\Delta_U^4 = U_{01} + U_{04} - (U_{p2} + U_{p3})$, $\Delta_D^4 = D_{p1} + D_{p4} - (D_{02} + D_{03})$		

Figure 9. Some types of differences among data subsets that can be formed with multiple cell setups. U and D refer to beam upstream or downstream. The subscripts "p" and "0" indicate the polarization state of the cell in a given period ("p"olarized or unpolarized - "0"). $\mathcal{A}$ is the acceptance, p_f is the packing fraction, Q is the beam charge and ε the detector efficiency.

4.1. A brief look at Systematic Errors

Systematic errors (other than scale factors) for a single difference of yields between two in-line cells depend on the relative differences in acceptances $\Delta\mathcal{A}$ and packing fractions Δp_f. Controlling those errors depends, in turn, on the uncertainties in those differences. Control can be accomplished by developing better ways of measuring p_f, and improving the understanding of relative acceptances by carrying out extensive simulations, using optics targets, etc.

The difference of yields from an unpolarized (U) and a longitudinally polarized ($\|$) cell depends on their respective acceptances, lengths ℓ, and the cross sections σ that apply in each case.

$$\Delta Y = \mathcal{A}^U \ell^U \sigma^U - \mathcal{A}^{\|} \ell^{\|} . \sigma^{\|} \tag{3}$$

Using the notation $\ell_A = N_0 z_A(\rho_A/M_A) = N_0 z_A \mathcal{D}_A$, where N_0 represents Avogadro's number, and ρ_A, M_A and z_A are the density, atomic or molecular mass, and thickness of nuclear species A, respectively, the unpolarized and polarized $\ell\sigma$ products are

$$\ell^U \sigma^U = N_0 z\Big(\mathcal{D}_{ND3}(\sigma_N^U + 3\sigma_D^U)p_f^U + \mathcal{D}_{He}\sigma_{He}^U(1 - p_f^U)\Big) \tag{4}$$

$$\ell^{\|} \sigma^{\|} = N_0 z\Big(\mathcal{D}_{ND3}(\sigma_N^U + 3(\sigma_D^U - \frac{Kx}{3} P_{zz} b_1))p_f^{\|} + \mathcal{D}_{He}\sigma_{He}^U(1 - p_f^{\|})\Big). \tag{5}$$

Replacing $p_f^{\|} = p_f^U - \Delta p_f$ and $\mathcal{A}^{\|} = \mathcal{A}^U - \Delta\mathcal{A}$, the difference can be written to first order as

$$\frac{\Delta Y}{s\mathcal{A}^U} - f\frac{\Delta p_f}{p_f^U} - \Big(f + \frac{\mathcal{D}_{He}}{\mathcal{D}_{ND3}}\frac{\sigma_{He}^U}{p_f^U}\Big)\frac{\Delta\mathcal{A}}{\mathcal{A}^U} = KxP_{zz}b_1(1 - \frac{\Delta p_f}{p_f^U} - \frac{\Delta\mathcal{A}}{\mathcal{A}^U}),$$

where $s = N_0 z p_f^U \mathcal{D}_{ND3}$ and $f = \sigma_N + 3\sigma_D - \sigma_{He}(\mathcal{D}_{He}/\mathcal{D}_{ND3})$.

Using this equation, some rough estimates can be made. For a solid angle difference at the limit of the spectrometer angular resolutions $\delta\theta$ and $\delta\phi$, $\partial\Delta\theta = \partial(\theta^U - \theta^{\|}) = \sqrt{2}\delta\theta$ and $\partial\Delta\phi = \sqrt{2}\delta\phi$, the relative difference in acceptances is the relative solid angle error, with

Journal of Physics: Conference Series **543** (2014) 012013 doi:10.1088/1742-6596/543/1/012013

$\partial(\Delta\Omega) = \partial(\Delta\theta\Delta\phi)$. Using the nominal HMS angle resolutions $\delta\theta = \delta\phi = 0.8$ mr the relative error on $\mathcal{A}$ is

$$\frac{\delta\Delta\mathcal{A}}{\mathcal{A}} = \frac{\delta\Delta\Omega}{\Delta\Omega} = \frac{2\delta\theta\delta\phi}{\Delta\Omega} = 5\times10^{-4}, \tag{6}$$

where an effective solid angle of 2.5 msr after acceptance matching cuts was used for this estimate (the full HMS solid angle is > 6 msr).

Approximating the nuclear cross sections as multiples of the deuteron cross section, $\sigma_N \simeq 7\sigma_D$, $\sigma_{He} \simeq 2\sigma_D$ so $f \sim 8.5\sigma_D$, the term in eq. (6) proportional to the relative difference of acceptances can then be compared to the r.h.s., which reduces to $P_{zz}b_1/F_1^D$, since $\sigma_D^U = KxF_1^D$. With $\mathcal{D}_{He}/\mathcal{D}_{ND3} = 0.8$, and using conservative values for $P_{zz} = 20\%$ and $p_f = 0.5$, the r.h.s is about 3.3×10^{-3} for $b_1 \sim 0.4\times10^{-2}$, as measured by HERMES [12]. The left side is $11.7\Delta\mathcal{A}/\mathcal{A} \sim 6\times10^{-3}$. Improvements in P_{zz}, packing fraction, and acceptance matching and resolution can reduce the l.h.s and increase the r.h.s to allow meaningful measurements.

If both cells could polarize the material in opposite tensor polarizations simultaneously, then an extra term appears inside the parenthesis of the r.h.s. of eq. (6), representing the ratio $p = |P_{zz}^+/P_{zz}^-| \simeq 1$ of the polarizations $P_{zz}^{+,-}$ of the upstream and downstream cells ($p = 0$ if only one cell is polarized), so an extra factor of about two would be gained for additional suppression of systematic effects.

To determine the contribution of the packing fraction difference to the systematic errors, we would consider the ammonia mass of about 9 g in the 3 cm long cells at 0.6 packing fraction. Careful weighing of polarizable material loads could resolve 1 mg in about 10 g, keeping this contribution under control. Material preparation in known shapes and dimensions might do even better than that.

In summary: Multiple cell polarized targets are being used and have been used very effectively for controlling systematic uncertainties in experiments with polarized targets, when it's difficult or impossible to flip helicities quickly and frequently, e.g. when using LiD. The in-line configuration favors using a simple solenoidal field for longitudinal spin observables, but transverse observables could still be accessible, at a price. This configuration could be a good option for a program on longitudinal vector and tensor spin observables at JLab. The study of new observables ten to a hundred times smaller than the traditional spin dependent ones requires advanced methods, like the one explored in this report. The potential use of multiple cell targets represents interesting target physics R&D in addition to its compelling physics motivation.

References

[1] Arenhovel H, Leidemann W and Tomusiak E 1988 *Z.Phys.* **A331** 123–138
[2] Leidemann W, Tomusiak E and Arenhovel H 1991 *Phys.Rev.* **C43** 1022–1037
[3] Hoodbhoy P, Jaffe R and Manohar A 1989 *Nucl.Phys.* **B312** 571
[4] Piegaia R N 1988 *UMI-89-17716, fig. 2.8* Ph.D. thesis
[5] Adams D *et al.* (Spin Muon Collaboration) 1999 *Nucl.Instrum.Meth.* **A437** 23–67
[6] `http://wwwcompass.cern.ch/compass/detector/target/Drawings/NH3target07v01.png`
[7] Koivuniemi J 2006 Status of the polarized COMPASS target, talk at Polarized Nucleon Targets for Europe, Rech
[8] `http://wwwcompass.cern.ch/compass/detector/target/Talks/TIS_meet_oct_00.pdf`
[9] Takabayashi N 2002 *CERN-THESIS-2002-051, fig. 5.1* Ph.D. thesis
[10] Richter A 2010 *CERN-THESIS-2010-266, fig. 5.7* Ph.D. thesis
[11] Levorato S 2010 *CERN-THESIS-2008-179, fig 6.4* Ph.D. thesis
[12] Airapetian A *et al.* (HERMES Collaboration) 2005 *Phys.Rev.Lett.* **95** 242001

Journal of Physics: Conference Series **543** (2014) 012014 doi:10.1088/1742-6596/543/1/012014

Tensor Target Polarization at TRIUMF

G Smith

Jefferson Lab, 12000 Jefferson Ave., Newport News, VA 23606 USA

E-mail: smithg@jlab.org

Abstract. The first measurements of tensor observables in $\pi\vec{d}$ scattering experiments were performed in the mid-80's at TRIUMF, and later at SIN/PSI. The full suite of tensor observables accessible in $\pi\vec{d}$ elastic scattering were measured: T_{20}, T_{21}, and T_{22}. The vector analyzing power iT_{11} was also measured. These results led to a better understanding of the three-body theory used to describe this reaction. A direct measurement of the target tensor polarization was also made independent of the usual NMR techniques by exploiting the (nearly) model-independent result for the tensor analyzing power at 90°_{cm} in the $\pi\vec{d} \to 2p$ reaction. This method was also used to check efforts to enhance the tensor polarization by RF burning of the NMR spectrum. A brief description of the methods developed to measure and analyze these experiments is provided.

1. Introduction

This article relies on the "Madison convention" [1]. Target vector and tensor polarizations are denoted by p_z and p_{zz} respectively. The tensor polarization of a recoil deuteron from πd elastic scattering measured in a double scattering experiment from an unpolarized target using a polarimeter is denoted t_{20}. The vector (iT_{11}) and tensor analyzing powers (T_{20}, T_{21}, and T_{22}) are the observables measured in a $\pi\vec{d}$ single scattering experiment from a polarized deuteron target (without a polarimeter). The composite observables τ_{21} and τ_{22} are defined below.

The starting expression for the tensor analyzing powers is

$$\frac{\sigma_{pol}}{\sigma_{unpol}} = (1 + a_{11} iT_{11} + a_{20}T_{20} + a_{21}T_{21} + a_{22}T_{20}) \tag{1}$$

where

$$a_{11} = \sqrt{3}p_z \sin(\alpha)\cos(\beta) \tag{2}$$

$$a_{20} = \frac{p_{zz}}{\sqrt{2}} \frac{3\cos^2(\alpha) - 1}{2} \tag{3}$$

$$a_{21} = \sqrt{3}p_{zz} \sin(\alpha)\cos(\alpha)\sin(\beta) \tag{4}$$

$$a_{22} = -\frac{\sqrt{3}}{2}p_{zz} \sin^2(\alpha)\cos(2\beta). \tag{5}$$

The Euler angle α describes the angle between the incoming beam axis (z) and the target magnetic field $\vec{B}$. The Euler angle β describes the angle between the y-axis ($\vec{k} \times \vec{k}'$) and the

Journal of Physics: Conference Series **543** (2014) 012014 doi:10.1088/1742-6596/543/1/012014

projection of $\vec{B}$ on the x-y plane. σ_{pol} and σ_{unpol} refer to the scattering cross sections measured with the target polarized or unpolarized.

To measure T_{20}, the target magnetic field is aligned along the incident beam axis ($\alpha = 0$, β undefined) to eliminate the contributions from the other T_{ij}. Then

$$T_{20} = \frac{\sqrt{2}}{p_{zz}} \left[\frac{\sigma_{pol}}{\sigma_{unpol}} - 1 \right]. \tag{6}$$

This was first measured in [2]. More results appeared in [3] and [4].

With $\alpha = 54.7°$ to eliminate T_{20}, and $\beta = 90°$ to eliminate iT_{11}, an observable is obtained sensitive mostly to T_{21} but with some admixture from T_{22}. In practice, due to the split coil geometry of the target magnet, α had to be chosen to be 45°, which also introduces a small T_{20} admixture. Then the composite observable

$$\tau_{21} \equiv T_{21} + \frac{1}{2} \left[\frac{T_{20}}{\sqrt{6}} + T_{22} \right] = \frac{2}{\sqrt{3} p_{zz}} \left(\frac{\sigma_{pol}}{\sigma_{unpol}} - 1 \right) \tag{7}$$

is used to access T_{21}. This was first measured in [5], and later in [3] and [4].

Finally, with $\alpha = 90°$ to eliminate T_{21}, and $\beta = 0°$ to maximize iT_{11} and T_{22} (but with some T_{20} admixture), the composite observable

$$\tau_{22} \equiv T_{22} + \frac{T_{20}}{\sqrt{6}} = \frac{2}{\sqrt{3} p_{zz}} \left(\frac{\sigma_{pol}}{\sigma_{unpol}} - 1 \right) \tag{8}$$

can be defined. τ_{22} was first measured in [6]. A measurement of the full suite of observables T_{20}, τ_{21}, and τ_{22} can be used to separate out the individual T_{ij}.

2. NMR Measurement of p_{zz}

The target tensor polarization p_{zz} can be determined from knowledge of the vector polarization p_z according to the usual expression

$$p_{zz} = 2 - \sqrt{4 - 3p_z^2}, \tag{9}$$

which is easily derived from the Brillouin formula with I=1. Thus the relative uncertainty in the tensor polarization $\Delta p_{zz}/p_{zz}$ is about twice the relative uncertainty in the vector polarization. The vector polarization can be determined either from comparison of the dynamically polarized NMR signal area A(dyn) to the thermal equilibrium NMR signal area A(TE), or from analysis of the NMR peak asymmetry [7]. In the former case, the thermal equilibrium polarization at temperature T and magnetic field B is

$$p_z(\mathrm{TE}) = \frac{4 \tanh x}{3 + \tanh^2 x} \simeq \frac{2\mu B}{3kT} \tag{10}$$

where $x = (\mu B)/(2kT)$, $\mu = 2.703 \times 10^{-14}$ MeV/T is the deuteron magnetic moment, and k is Boltzmann's constant. For 0.5 K and 2.5 T, p_z(TE)=0.00105. Then,

$$p_z(\mathrm{dyn}) = \frac{A(\mathrm{dyn})}{A(\mathrm{TE})} p_z(\mathrm{TE}). \tag{11}$$

In the latter case, with $R = \exp{(-h\nu_D)}/(kT_S)$ denoting the ratio of the two NMR signal peaks (which are split by the hyperfine interaction of the deuteron quadrupole moment with the electric

Journal of Physics: Conference Series **543** (2014) 012014 doi:10.1088/1742-6596/543/1/012014

field gradients associated with the two different molecular bonds), ν_D the deuteron NMR signal frequency (16.6 MHz at 2.5 T), and T_S the deuteron spin temperature,

$$p_z = \frac{1 - R^2}{1 + R + R^2}. \tag{12}$$

To achieve even modest tensor polarizations of 0.20, vector polarizations of 0.50 must be achieved. For the deuterated butanol and propandiol targets used in the $\pi\vec{d}$ scattering programs at TRIUMF and SIN, this required migrating from ^{3}He refrigerators with characteristic temperatures of ~0.5° K to ^{3}He/^{4}He dilution refrigerators with characteristic temperatures of ~0.05° K. Values of p_z~0.48 (p_{zz}~0.18) were routinely achieved in [3], [4], and [6].

3. Benchmarking NMR methods

Determining the tensor polarization from the vector polarization assumes the quadrupole moment contribution is negligible, that the distribution of spins can be described by a Boltzmann distribution (the equal spin temperature hypothesis), and that the NMR system used to measure p_z is linear over a wide range. For the first measurements in the $\pi\vec{d}$ program which relied on a tensor polarized target [2], the target tensor polarization determined with the NMR methods described above was verified by direct measurement, using a technique which did not rely at all on NMR signal methods. This was accomplished [8] by recognizing that at a center of mass angle of 90°, the tensor analyzing power T_{20} in the $\pi\vec{d} \to 2p$ reaction is maximal ($-\sqrt{2}$) and that this is a virtually model-independent result. Therefore,

$$p_{zz} = \frac{\sqrt{2}}{T_{20}} \left(\frac{\sigma_{pol}}{\sigma_{unpol}} - 1 \right), \tag{13}$$

where obviously the polarized and unpolarized cross sections refer to the measurement of $\pi\vec{d} \to 2p$ at $90^\circ_{c.m.}$.

T_{20} in this expression was determined three independent ways. First, use was made of a published measurement of A_{yy} at 90° in the $pp \to \pi d$ reaction [9], which yielded $A_{yy} = -0.86 \pm 0.04$ at T_p=447 MeV. Then

$$T_{20} = \sqrt{2}\, \frac{3A_{yy} - 1}{4} = -1.27 \pm 0.05. \tag{14}$$

Alternatively, use was made of the results of partial-wave analysis. In the limit that a_2 (which feeds the $^1D_2$ pp partial-wave) dominates, as expected on the (3,3) resonance,

$$T_{20} = \frac{-\sqrt{2}}{70} \frac{(70 + 25P_2(\cos\theta) + 108P_4(\cos\theta))}{1 + P_2(\cos\theta)} = -\sqrt{2} \text{ at } 90°. \tag{15}$$

Three-body Fadeev equations also provide predictions. Combining all this information together, and accounting for the ±2.5° angular acceptance in the experiment performed to measure p_{zz}, a value of $T_{20} = -1.28 \pm 0.03$ was arrived at and used to determine the target tensor polarization p_{zz} via Eq. 13 with measurements of σ_{pol} and σ_{unpol} in pion absorption at $90^\circ_{c..m.}$.

The $\pi\vec{d} \to 2p$ measurement was performed [8] at TRIUMF using a 95% deuterated butanol target (C_4D_9OD) and a non-deuterated butanol target (C_4H_9OH) for background subtraction. The latter constitutes a perfect background target of the same size (~1 mm) butanol spheres in the same 16x16x5 mm^3 teflon basket immersed in the same ^{3}He/^{4}He mixture, without the possibility of absorption on the hydrogen that substitutes for the deuterium. The superconducting split-pair solenoid was oriented with the magnetic field along the beam

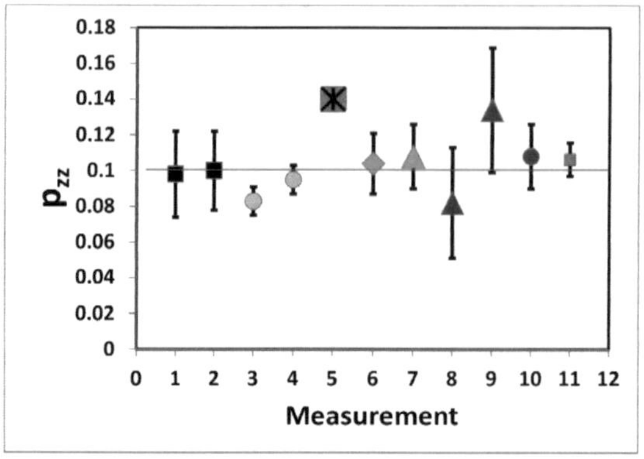

Figure 1. Summary of RF burning results from [8]. Measurements 1&2 were obtained by direct measurement before burning (see text) using TOF and co-planarity data, respectively. NMR measurements 3 & 4 correspond to the peak ratio (Eq. 12) and area (Eq. 11) methods, respectively. Measurement 5 was estimated in [11] from analysis of the NMR signal after RF burning; no uncertainty was provided. Direct measurements 6 & 7 are the analogs of 1 & 2 after burning. Measurements 6 & 7 were repeated and are shown as measurements 8 & 9. Measurement 10 was also obtained after burning, but at 1.25 T instead of the 2.5 T used for all the other measurements. The weighted mean of all the direct measurements after burning (6–10) is shown as measurement 11.

direction. An opening angle of $\pm 12.5^\circ$ about 90° between the split-pair coils provided sufficient acceptance for the detection of the reaction products in coincidence. Two arms consisting of a pair of wire chambers and scintillation counters in each arm were used to make the measurement using both time-of-flight and co-planarity information.

The time-of-flight analysis yielded $p_{zz} = 0.098 \pm 0.024$ and the co-planarity method yielded $p_{zz} = 0.100 \pm 0.022$. This compared well with the more precise NMR information: $p_{zz} = 0.083 \pm 0.008$ using the ratio of dynamic and thermal equilibrium NMR signal areas, and $p_{zz} = 0.095 \pm 0.008$ using the NMR peak area ratio method of Eq. 12. This measurement established within its uncertainty that the NMR methods are valid for determining the tensor polarization of a target.

3.1. RF Burning

While the benchmarking setup described above was in place, measurements were also performed to study the potential enhancement of the target tensor polarization from RF burning the NMR spectrum. RF burning distorts the NMR signal, and it has been suggested [10] this is the result of dramatic enhancement in tensor polarization. However, it is difficult to know how much enhancement has taken place (if any), or what decay time is associated with the potentially enhanced signal. The only way to quantitatively test this is to avoid reliance on the usual NMR techniques, and instead determine the target polarization using the direct measurement technique described above in Sec. 3.

The NMR signal was burned [11] by applying RF to one pedestal until one peak was completely eliminated. From NMR signal analysis it was estimated that the tensor polarization had been increased by ~ 0.05, *i.e.* from ~ 0.09 before burning to ~ 0.14 after burning.

Journal of Physics: Conference Series **543** (2014) 012014 doi:10.1088/1742-6596/543/1/012014

Measurements of the $\pi\vec{d} \to 2p$ reaction were performed as described in Sec. 3 over a period of 18 hours after burning, with the target in frozen spin mode (no μ-waves) at 2.5 T.

The results were disappointing. The TOF technique yielded $p_{zz} = 0.10 \pm 0.017$ and the coplanarity technique yielded $p_{zz} = 0.11 \pm 0.018$, an enhancement relative to the unburned TOF/coplanarity results of only $p_{zz}(\text{burned})/\text{p}_{\text{zz}}(\text{unburned}) = 1.08 \pm 0.3$. Within the rather large uncertainties of the measurement, no significant enhancement from RF burning was observed. Similar results were obtained with a holding field of 1.25 T. Shorter time periods, or higher holding fields were not explored. The results are summarized in Fig. 1. The weighted mean of all five direct measurements performed after burning was 0.106±0.009.

4. Analysis Methods

Equations 6, 7, and 8 were used to determine the tensor observables in $\pi\vec{d}$ elastic scattering. Unpolarized cross sections were measured, as well as polarized cross sections with only one sign of target polarization. Other analysis methods were also explored which produce equivalent results, but probe potential sources of systematic error differently.

One variation was explored in [4] at SIN (now PSI). It consists of going back to Eq. 1 and recasting it in the form

$$\sigma(p_z) = A + BP_z + Cp_{zz} \tag{16}$$

where $A = \sigma^0$, $B = \sigma^0 a_V iT_{11}$, & $C = \sigma^0 a_T T$, and σ^0 refers to the unpolarized cross section. For the case of $T = T_{20}$, $a_T = 1/\sqrt{2}$. For the case of $T = \tau_{21}$, $a_T = \sqrt{3}/2$. Several measurements with positive, negative, and zero p_z provide a body of data to fit the coefficients A, B and C. If $B \neq 0$, then either there is a problem with the measurement of p_z or else the target magnetic field is misaligned with respect to the beam axis ($\alpha \neq 0$).

Another analysis variant was also explored [4] at SIN; a matrix method. Positive and negative vector polarization data were acquired interspersed with unpolarized data, as in the sequence $..., \sigma^+), (\sigma^+, \sigma^{0+}), (\sigma^{0-}, \sigma^-), (\sigma^-, ...$ Writing

$$\sigma^{\pm} = \sigma^{0\pm}\left(1 \pm a_V p_z^{\pm} iT_{11} + a_T p_{zz}^{\pm} T\right), \tag{17}$$

and adding and subtracting, one finds

$$a_T T = \frac{p_z^-(\sigma^+/\sigma^{0+}) + p_z^+(\sigma^-/\sigma^{0-}) - (p_z^+ + p_z^-)}{p_z^+ p_{zz}^- + p_z^- p_{zz}^+} \tag{18}$$

and similarly for $a_V iT_{11}$. Then matrices of $a_T T$ can be constructed for each pair, as in (σ^+, σ^{0+}). With N pairs of (σ^+, σ^{0+}) and another N pairs of (σ^-, σ^{0-}), an NxN matrix can be formed whose diagonal elements are the time-ordered pairs. The weighted average of the diagonal elements is the result. Electronic drifts average out. The individual row and column averages provide a measure of the consistency of the data.

5. The Physics of the $\pi\vec{d}$ Experiments

Initially, interest in measuring the $\pi\vec{d}$ tensor analyzing power T_{20} was motivated by an experiment [12] which used an unpolarized target and a ^{3}He$(\vec{d}, p)^4$He polarimeter to measure the tensor polarization T_{20} of the recoil deuteron in πd elastic scattering. Large oscillations in the observed angular distribution at an incident energy of 134 MeV were absent at neighboring energies as close as ±15 MeV away. The team which published these results claimed they constituted evidence for a narrow dibaryon resonance (a six-quark state of matter). Similar experiments were performed at LAMPF [13] and at TRIUMF [14] using the same technique (a ^{3}He recoil deuteron polarimeter in a double scattering experiment with an unpolarized d target) but no such evidence was found. The original group from SIN repeated their experiment [15] with a redesigned polarimeter, confirming and strengthening their original conclusions.

Given this impasse on such a potentially important finding, at TRIUMF a program was initiated [2] to measure the tensor analyzing power T_{20} with a tensor polarized target in a single scattering experiment, to avoid the same systematic errors which obviously plagued at least one of the three t_{20} groups employing recoil polarimeters in double scattering experiments. If a tensor polarized target could be used in a single scattering $\pi\vec{d}$ experiment, the measurements would have much better statistics and would be relatively free of systematic errors, at least free of the same systematic errors which had to be dealt with in the polarimeter experiments.

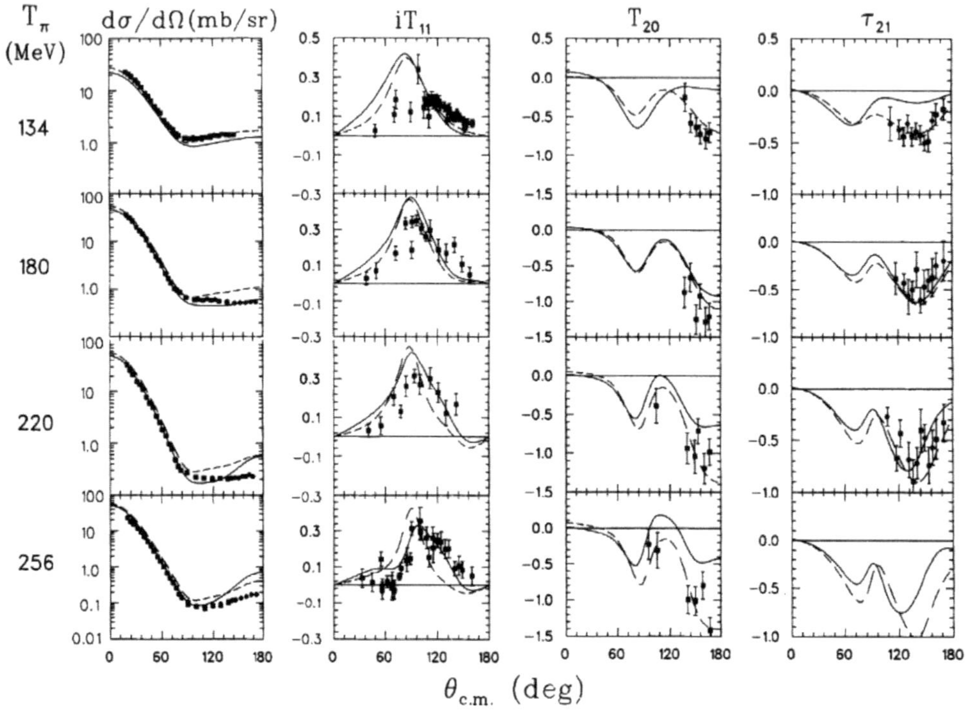

Figure 2. Measurements of πd elastic scattering observables in the (3,3) resonance region from [3]. The differential cross-sections are from [17] and [18]. The iT_{11} data are from [19], the T_{20} data are from [2] and [3], and the τ_{21} data are from [5] and [3]. The three-body calculations [20] show the predictions with (solid curves) and without (dashed curves) the πN P_{11} contribution.

A simple detector system was built employing scintillation counters for the coincident detection of the pion and recoil deuteron. The system provided a large solid angle (180 msr), and six-fold angle multiplicity. The time-of-flight (TOF) and pulse heights of the pion and deuteron were recorded. The recoil deuteron arms were composed of thin and thick deuteron counters sandwiching an aluminum absorber whose thickness was tailored to range out recoil deuterons, followed by a veto counter. The pion arms were each composed of a two-element scintillation counter telescope. The incident beam was directly counted in a coincidence of two thin in-beam

Journal of Physics: Conference Series **543** (2014) 012014 doi:10.1088/1742-6596/543/1/012014

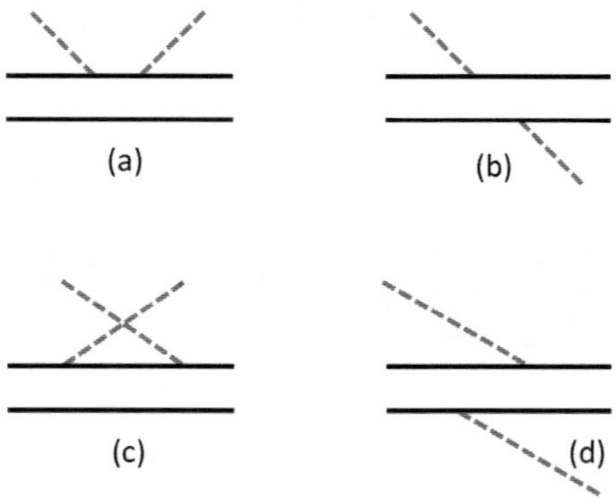

Figure 3. Examples of the standard pole (a) and non-pole (b) diagrams for the P_{11} πN partial wave. The corresponding crossed diagrams are shown in (c) and (d), respectively.

scintillation counters, and a split counter hodoscope was located downstream of the target. The latter was particularly useful in achieving the crucial alignment of the incident pion momentum with the target magnetic field.

The tensor analyzing power T_{20} and the recoil deuteron tensor polarization t_{20} can be related to each other [16] with the Wigner d functions $d^{i}_{jk}(\theta_d)$ according to

$$t^{\rm lab}_{20}(\theta_d) = T^{\rm c.m.}_{20}(\theta_d)d^2_{00}(\theta_d) - 2T^{\rm c.m.}_{21}(\theta_d)d^2_{10}(\theta_d) + 2T^{\rm c.m.}_{22}(\theta_d)d^2_{20}(\theta_d). \tag{19}$$

The rotation accounts for the fact that in the t_{20} measurements, the z-axis points along the recoil deuteron momentum, but in the T_{20} measurements, it points along the incident beam direction. A further change from $t^{\rm lab}_{20}$ to $t^{\rm c.m.}_{20}$ admixes in T_{21} and T_{22} components, which are small in the relevant angular range. The results of the tensor analyzing power measurements found no dibaryon oscillations in the measured angular distributions, and were consistent with the t_{20} results from [13] and [14], settling this particular dibaryon issue once and for all.

Comparison of the T_{20} measurements with 3-body Fadeev calculations showed that our understanding of the results was incomplete, however. To try and shed light on this aspect of the results, measurements of the other two tensor observables in $\pi\vec{d}$ scattering were performed. These also showed the puzzling result that calculations performed without the P_{11} πN partial waves described the results better than the calculations which included this term. The effect was especially pronounced in the T_{20} data, as shown in Fig. 2.

This 3-body problem was finally resolved when Jennings pointed out [21] that some of the diagrams (see Fig. 3) which contribute to this partial wave were not being included in the three-body Fadeev calculations. The nucleon pole (non-pole) terms are diagrams in which the pion is absorbed on and re-emitted from the same (different) nucleon. The non-pole terms are required to satisfy the Pauli principle in the intermediate state. The pole and non-pole contributions are large, of similar magnitude, but opposite sign and thus mostly cancel to give the small P_{11} πN phase shift.

For each of the two pole and two non-pole diagrams, there are differently time-ordered counterparts in which the pion legs are crossed instead of uncrossed. The crossed diagrams are essentially four-body in nature. Although both the crossed and uncrossed pole terms are included naturally as part of the P_{11} πN amplitude, the non-pole terms are not true pion absorption, but multiple scattering. Thus the crossed non-pole terms are not included, and since they are explicitly four-body diagrams they cannot be treated by the three-body theory.

Jennings showed however that the omitted non-pole crossed diagrams would essentially cancel the contributions from the non-pole uncrossed diagrams which were included, thus explaining why the calculations without the P_{11} partial wave described the data better than the calculations with (only part of) this partial wave.

6. Summary

The first measurements of tensor observables were obtained in the $\pi\vec{d}$ system using tensor polarized targets of deuterated butanol and propandiol. After migration to dilution refrigerators, target vector polarizations of $p_{zz} = 0.48$ were routinely obtained, corresponding to tensor polarizations $p_{zz} = 0.18$. In pion beams of only a few MHz, the cooling powers required were typically only 3–4 mW. The polarizing fields were 2.5 T for these measurements, and temperatures of 50-70 mK were typical with the dilution refrigerators.

The impact of RF burning the NMR signal on the target polarization was explored. By definition this technique radically distorts the NMR signal, making polarization measurement problematic via the usual NMR techniques. Therefore, the target polarization after burning was determined without using the NMR signal at all. Instead, the known tensor analyzing power in pion absorption was exploited to measure p_{zz} before and after RF burning. Unfortunately, no significant enhancement of the polarization was observed within the uncertainties after RF burning. The uncertainties were large enough, however, that some degree of enhancement cannot be ruled out. Higher polarizing fields, lower temperatures, and different target materials were not explored in this study, although they may have led to better results. One conclusion from this work is that independent techniques like the direct measurement described above are an absolute necessity to provide realistic target polarizations after RF burning. Reliance on NMR methods alone was confirmed for unburned targets, but is not an option for targets that have been RF burned in an attempt to enhance the polarization.

The measurement of the tensor observables in $\pi\vec{d}$ elastic scattering completed a program that started by mapping the differential cross section and vector analyzing power over the (3,3) πN resonance region. Techniques used to acquire and analyze the results of the tensor analyzing powers were presented. The measurements led to a better understanding of the three-body Fadeev calculations that were compared to the data. In particular the treatment of the P_{11} πN partial wave and the role of pion absorption in the theory were better understood as a result of these measurements.

7. Acknowledgements

I would like to thank all of my collaborators in the $\pi\vec{d}$ tensor programs at TRIUMF and SIN. This work is dedicated to Erich Vogt, who was director of TRIUMF from 1981–1994, during the years this research was carried out, and who passed away earlier this year. The stimulating and productive environment of scientific freedom and encouragement he fostered among the staff and users was a huge factor in the success of the program described here, as well as many others at TRIUMF.

The research described in this article was supported by TRIUMF and by grants from the Natural Sciences and Engineering Research Council of Canada (NSERC). The work preparing this article was supported by DOE contract No. DE-AC05-06OR23177, under which Jefferson Science Associates, LLC operates Thomas Jefferson National Accelerator Facility.

References

[1] Schwandt P and Haeberli W 1968 *Nucl. Phys.* **A110** 585
[2] Smith G R, *et al* 1986 *Phys. Rev. Lett.* **57** 803
[3] Smith G R, *et al* 1988 *Phys. Rev. C* **38** 251
[4] Ottermann C R *et al* 1988 *Phys. Rev. C* **38** 2296

[5] Smith G R *et al* 1987 *Phys. Rev. C* **35** 2343
[6] Ottermann C R *et al* 1988 *Phys. Rev. C* **38** 2310
[7] Hamada O *et al* 1981 *Nucl. Instrum. Methods* **A189** 561
[8] Smith G R *et al* 1987 *Nucl. Instrum. Methods* **A254** 263
[9] Aprile-Giboni E *et al* 1984 *Nucl. Phys. A* **415** 391
[10] De Boer W *et al* 1974 *Phys. Lett. A* **46** 143
[11] Delheij P, Healey D, and Wait G D, TRIUMF report TRI-PP-86-027
[12] Ulbricht J *et al* 1982 *Phys. Rev. Lett.* **48** 311; Gruebler W *et al* 1982 *Phys. Rev. Lett.* **49** 444; Konig V *et al* 1983 *J. Phys. G* **9** L211
[13] Holt R J *et al* 1979 *Phys. Rev. Lett.* **43** 1229; Holt R J *et al* 1981 *Phys. Rev. Lett.* **47** 472; Ungricht E *et al* 1984 *Phys. Rev. Lett.* **52** 333; Ungricht E *et al* 1985 *Phys. Rev. C* **31** 934
[14] Shin Y M *et al* 1985 *Phys. Rev. Lett.* **55** 2672
[15] Swiss Institute for Nuclear Research Annual Report No. NL18, 1984 (unpublished).
[16] Grein W and Locher M P 1981 *J. Phys. G* **7** 1355
[17] Gabathuler K *et al* 1980 *Nucl. Phys.* **A350** 253
[18] Ottermann C *et al* 1985 *Phys. Rev. C* **32** 928
[19] Smith G R *et al* 1984 *Phys. Rev. C* **29** 2206
[20] Blankleider B and Afnan I R 1981 *Phys. Rev. C* **24** 1572
[21] Jennings B R 1988 *Phys. Lett. B* **205** 187

Journal of Physics: Conference Series **543** (2014) 012015 doi:10.1088/1742-6596/543/1/012015

Investigation into tensor polarization enhancement of solid polarized targets

D Keller

University of Virginia, Physics Department, 382 McCormick Rd, Charlottesville, VA 22904-4714, USA

E-mail: dustin@jlab.org

Abstract. A discussion of tensor polarization of solid targets produced by Dynamic Nuclear Polarization and Dynamic Nuclear Orientation in the use of nuclear physics experiments is presented. Techniques in the deuteron tensor polarization enhancement and enhanced tensor polarization measurement uncertainty are also discussed in the interest and preparation of future experiments.

1. Introduction

In recent years there has been a rise in the interest of solid tensor polarized targets for the use in nuclear physics experiments. Research in optimization and measurement of tensor polarization is now currently underway at the University of Virginia's polarized target lab. Some methods and direction of that research is discussed.

For the explicit use in nuclear experiments the polarized target system is prepared so that the momentum and spin state of the deuteron can constitute the beam target. For the sake of optimizing luminosity of a tensor target, a new investigation has begun to maximize the desired helicity state in an ensemble comprised of a large number of deuterons present in a macroscopic solid target. What is expected to be achievable for use with a photon beam is considerably greater tensor enhancement of a target due to the frozen spin state available using a dilution fridge. The large current desired for electron beams makes the frozen spin state not a consideration at present. There is still the possibility to enhance the tensor polarization that is achieved with a $\sim$1 K evaporation fridge that can be used with an electron beam.

Naturally a minimization procedure of polarization errors [1] is always essential, but there will be additional uncertainties that pertain only to the tensor polarization. This is true no matter the technique of acquiring the tensor polarization.

2. Tensor Polarization at Equilibrium

The deuteron, a spin-1 target, has energy levels which split three ways in a magnetic field, $m = -1, 0, +1$. The target spin orientation can be described using the vector and tensor polarization. The tensor polarization, P_{zz}, can be expressed in terms of the vector polarization as $P_{zz} \doteq 2 - \sqrt{4 - 3P^2}$ where the dot implies the relation is true when thermal equilibrium exists within the deuteron spin species. The definition of vector polarization for spin-1 is,

$$P = \frac{n_+ - n_-}{n_+ + n_- + n_0} \doteq \frac{r^2 - 1}{r^2 + r + 1}, \tag{1}$$

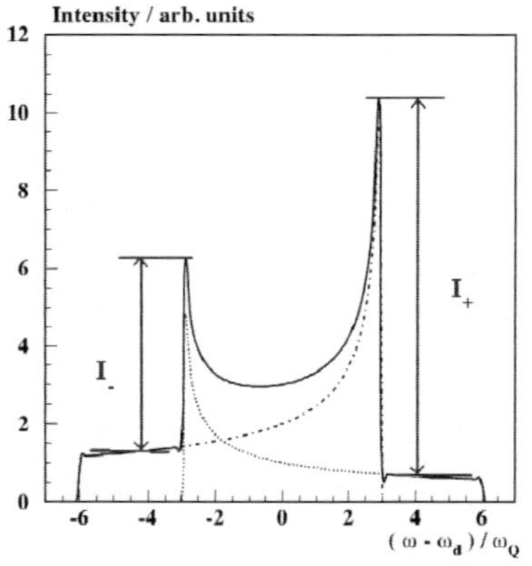

Figure 1. Deuteron magnetic resonance line shape and peak intensities I_+ and I_-.

with the tensor polarization defined as,

$$P_{zz} = \frac{n_+ - 2n_0 + n_-}{n_+ + n_- + n_0} \doteq \frac{r^2 - 2r + 1}{r^2 + r + 1}. \tag{2}$$

Where the second equality in each case is not part of the definition but a relation to the transition ratio r defined as $r = I_+/I_-$, see Figure 1. These relations in terms of $r = e^{\beta\hbar\omega_d}$ are true to first order in $\beta\hbar\omega_q$, where ω_d (ω_q) is Larmor (quadrapole interaction) frequency. The intensities I_- and I_+ can be determined using a fitting procedure [2]. The resulting polarization measurements can be used complementary to the TE signal polarization calibration. For vector polarizations over 30% the two independent methods result in a discrepancy smaller than 2.5%. This leads to a minimum uncertainty in the natural tensor polarization of approximately 5%. The relationship between the ratio of intensities and the absolute error with increasing polarization can be seen in Figure 2.

3. Tensor Polarization Enhancement

Tensor enhancement can be achieved in an inhomgeneously broadened DMR line by selectively saturating only some of its components. A homogeneously broadened line can saturate uniformly where the degree of saturation is dependent on the spectral density at that frequency and the power rate of the RF. The saturated part of the line will recover exponentially with a time constant. Cross relaxation has a central role in the deuteron alignment changes. Spatial inhomogeneity comes from local gradients in the external field. Different spin populations in different parts of the sample can come into equilibrium by spin diffusion in space. This is a slow process so RF saturation can readily occur. A perturbation at a single frequency or frequency range will recover uniformly in a time set by the cross relaxation rate and then return to equilibrium over a longer time interval defined by the time constant.

Deuteron spin alignment can be manipulated at a specific frequency when exposed to a modulated RF field using an external coil around the target cup, as seen in Figure 3. The accuracy and enhancement of the tensor polarization is greatly dependent on the polarization technique. For nuclear experiments with the use of an electron beam the focus is presently on positive enhanced tensor polarization which can be expressed as,

$$P_{zz} = C(A_+ - A_-). \tag{3}$$

Journal of Physics: Conference Series **543** (2014) 012015 doi:10.1088/1742-6596/543/1/012015

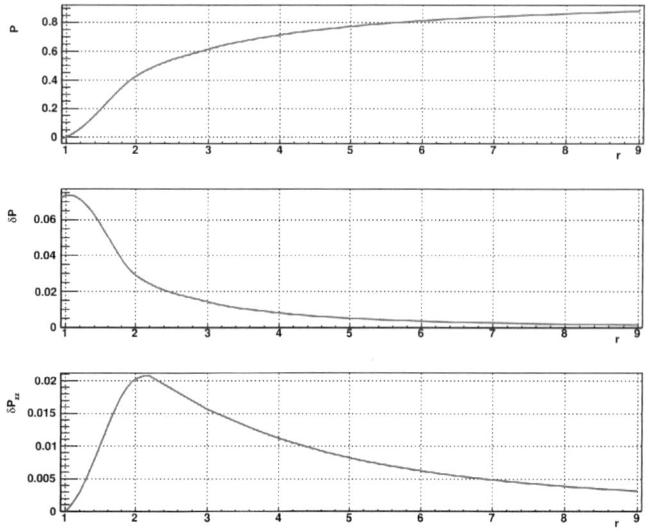

Figure 2. Polarization and absolute error with increasing r.

This is an expression that assumes the knowledge of the area of each of the two transition in the manipulated DMR line. Here C is the calibration constant. A positive tensor polarization enhancement occurs only when the $n_+ + n_-$ population increases with respect to n_0 population. To optimize the tensor polarization the vector polarization must be maximized using DNP at which point the RF-modulation can be used to distort the spectral density over time. The RF-modulation induces thermal cooling with frequency dependence by simultaneous flip-flops while also suppressing transitions between either the $m = -1 \rightarrow m = 0$ or $m = 0 \rightarrow m+ = 1$. A full saturation of the smaller transitions non-overlapping region (pedestal) can be especially helpful for enhancement. Tensor optimization will occur when the range in RF-modulation is chosen as to maximize the area of the resulting signal while minimizing the area in the overlap. The usefulness and associated uncertainty depends greatly on knowing the area of each of the two transitions under the RF manipulation. This has yet to be shown.

A solid tensor polarized target using both negative and positive tensor polarizations is plausible for use with a photon beam or low current electron beam. One way to achieve negative tensor polarization is to use dynamic nuclear orientation on a material that contains a separate proton spin-spin reservoir such as the -diol compounds. Applying a RF-modulation slightly off the Larmor frequency of the proton spins can lead to enhanced alignment of the deuteron spins independent of the vector polarization. The off-resonance RF field cools the proton spin-spin interaction reservoir and is transmitted to the deuteron quadrupole interaction reservoir [3]. The deuteron spin alignment increase can be calculated in much the same way with the accuracy purely dependent on the knowledge of the separate area of each peak.

Another very quick and simple technique is Adiabatic Fast Passage (AFP). This is also a very likely option for the frozen spin target. Spin-polarization reversal by using the adiabatic-fast passage (AFP) mechanism allows for quick target helicity reversal. This procedure is used frequently with gas targets (ie. Helium gas) but has yet to be employed for a solid polarized target in nuclear physics experiments.

The AFP process requires the nuclear-spin system to be placed in a strong magnetic field H_0 while being irradiated with a RF field H_1, linearly polarized in a plane perpendicular to H_0,

Journal of Physics: Conference Series **543** (2014) 012015 doi:10.1088/1742-6596/543/1/012015

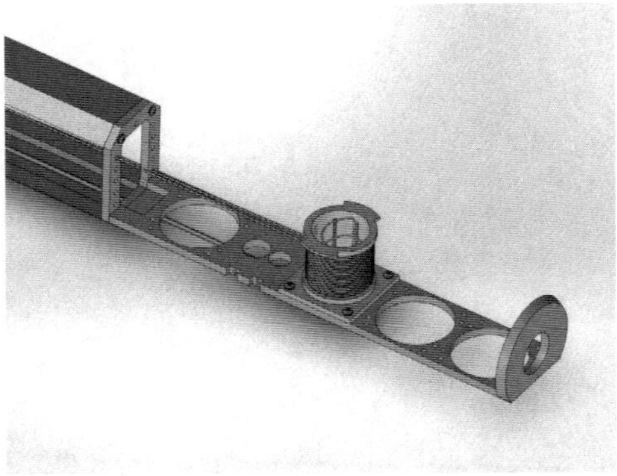

Figure 3. Target cup and coil used to generate the modulated RF field (Diagram by A. Rijal).

with frequency close to the NMR frequency. The passage is performed by varying the frequency ω. This can be understood as a transfer of order from the Zeeman system to the the quadrapole system for deuterons, then back to the Zeeman system with opposite spin orientation. The condition for adiabaticity is,

$$\frac{dH}{dt} \ll \gamma H_1^2,$$

where dH/dt is the sweep rate.

The quality of the coupling between these subsystems by the field H_1 along with the RF field characteristics and the sweep time τ lead to an efficiency which is defined by the loss in the magnitude of polarization.

Once the AFP procedure is well understood for solid polarized targets, the RF field characteristics and sweep time must be investigated for many materials so that the AFP efficiency can be correlated to optimal polarization and target performance at the required temperature.

4. Conclusion

Tensor enhancement can be achieved by keeping the system in equilibrium at a higher holding fields or perhaps using non-standard target materials, outside of this, enhancement must be achieved through other means. One possibility is with the use of RF to manipulate the alignment of the deuterons in the ensemble. Some plausible methods are mentioned, all of which require further research and development in the lab as well as research into the measurement techniques.

References

[1] Keller D 2013 *Nucl. Instr. and Meth.* A **728** 133
[2] Dulya C *et al.* 1997 *Nucl. Instr. and Meth.* A **398** 109
[3] de Boer W *et al.* 1973 *Phys.Lett.* B **46** 143

AUTHOR INDEX